T0311891

THE ROLE OF PUBLIC PARTICIPATION IN ENERGY TRANSITIONS

THE ROLE OF PUBLIC PARTICIPATION IN ENERGY TRANSITIONS

Edited by

ORTWIN RENN

Institute for Advanced Sustainability Studies (IASS), Potsdam, Germany

FRANK ULMER

Dialogik Non-profit Institute for Communication and Cooperation Research, Stuttgart, Germany

ANNA DECKERT

Dialogik Non-profit Institute for Communication and Cooperation Research, Stuttgart, Germany

Academic Press is an imprint of Elsevier
125 London Wall, London EC2Y 5AS, United Kingdom
525 B Street, Suite 1650, San Diego, CA 92101, United States
50 Hampshire Street, 5th Floor, Cambridge, MA 02139, United States
The Boulevard, Langford Lane, Kidlington, Oxford OX5 1GB, United Kingdom

British Library Cataloguing-in-Publication Data
A catalogue record for this book is available from the British Library

Library of Congress Cataloging-in-Publication Data
A catalog record for this book is available from the Library of Congress

ISBN: 978-0-12-819515-4

For Information on all Academic Press publications
visit our website at https://www.elsevier.com/books-and-journals

Publisher: Brian Romer
Acquisitions Editor: Graham Nisbet
Editorial Project Manager: Sara Valentino
Production Project Manager: Nirmala Arumugam
Cover Designer: Matthew Limbert

Typeset by MPS Limited, Chennai, India

Contents

6. Cosmopolitan governance for sustainable global energy transformation: democratic, participatory-deliberative, and multilayered

Andreas Klinke

Part 2
Case studies

7. The Kopernikus Project ENavi: linking science, business, and civil society

Piet Sellke, Matthias Bergmann, Marion Dreyer, Oskar Marg and Steffi Ober

8. Climate change policies designed by stakeholder and public participation

Jörg-Marco Hilpert and Oliver Scheel

9. Digital tools in stakeholder participation for the German Energy Transition. Can digital tools improve participation and its outcome?

Anna Deckert, Fabian Dembski, Frank Ulmer, Michael Ruddat and Uwe Wössner

10. Citizen participation for wind energy: experiences from Germany and beyond

Gundula Hübner

11. The contact group—public participation in the distribution network expansion in Baden-Württemberg

Regina Schroeter and Daniel Zirke

12. Social sustainability: making energy transitions fair to the people

Daniela Setton

13. Conclusion

Ortwin Renn, Anna Deckert and Frank Ulmer

List of contributors

Matthias Bergmann Institute for social-ecological Research, Frankfurt, Germany

Linda Connor University of Sydney, NSW, Australia

Anna Deckert Dialogik Non-profit Institute for Communication and cooperation research, Stuttgart, Germany

Fabian Dembski High Performance Computing Center, Stuttgart, Germany

Emily Drewing Department for Political Science, University of Siegen, Siegen, Germany

Marion Dreyer Dialogik Non-profit Institute for Communication and cooperation research, Stuttgart, Germany

Eva Eichenauer Research Department Institutional Change and Regional Public Goods, Leibniz Institute for Research on Society and Space (IRS), Erkner, Germany

Devleena Ghosh University of Technology, Faculty of Arts and Social Sciences, Sydney, Australia

Jörg-Marco Hilpert Dialogik Non-profit Institute for Communication and cooperation research, Stuttgart, Germany

Lars Holstenkamp Leuphana University of Lüneburg, Institute of Finance and Accounting, Lüneburg, Germany

Gundula Hübner MSH Medical School, Hamburg & Institut für Psychologie, Martin-Luther-Universität Halle-Wittenberg, Halle (Saale), Germany

Jan Warode Kulturwissenschaftliches Institut Essen (KWI), Institute for Advanced Study in the Humanities, Essen, Germany

Jan-Hendrik Kamlage Kulturwissenschaftliches Institut Essen (KWI), Institute for Advanced Study in the Humanities, Essen, Germany

Andreas Klinke Environmental Policy Institute, Memorial University of Newfoundland, Canada

Oskar Marg Institute for social-ecological Research, Frankfurt, Germany

Jonathan Paul Marshall University of Technology, Faculty of Arts and Social Sciences, Sydney, Australia

Franziska Mey University of Technology Sydney, Institute for Sustainable Futures, Ultimo, NSW, Australia

Tom Morton University of Technology, Faculty of Arts and Social Sciences, Sydney, Australia

Katja Müller Martin-Luther-Universität Halle-Wittenberg, Zentrum für Interdisziplinäre Regionalstudien (ZIRS), Germany

Steffi Ober Zivilgesellschaftliche Plattform Forschungswende, Berlin, Germany

Jörg Radtke Department for Political Science, University of Siegen, Siegen, Germany

Ortwin Renn Institute for Advanced Sustainability Studies (IASS), Berliner Strasse, Potsdam, Germany

Michael Ruddat Center for Interdisciplinary Risk and Innovation Studies (ZIRIUS), Stuttgart, Germany

Oliver Scheel Center for Interdisciplinary Risk and Innovation Studies (ZIRIUS), Stuttgart, Germany

Regina Schroeter Netze BW, Stuttgart, Germany

Pia-Johanna Schweizer Institute for Advanced Sustainability Studies (IASS), Berliner Strasse, Potsdam, Germany

Piet Sellke Dialogik Non-profit Institute for Communication and cooperation research, Stuttgart, Germany

Daniela Setton Institute for Advanced Sustainability Studies (IASS), Berliner Strasse Potsdam, Germany

Frank Ulmer Dialogik Non-profit Institute for Communication and cooperation research, Stuttgart, Germany

Jana Wegener Kulturwissenschaftliches Institut Essen (KWI), Institute for Advanced Study in the Humanities, Essen, Germany

Uwe Wössner High Performance Computing Center, Stuttgart, Germany

Daniel Zirke Netze BW, Stuttgart, Germany

About the editors

Ortwin Renn is the Scientific Director at the International Institute for Advanced Sustainability Studies (IASS) in Potsdam (Germany). He serves as full professor for environmental sociology and technology assessment at the University of Stuttgart. He also directs the nonprofit company DIALOGIK, a research institute for the investigation of communication and participation processes. He is an Adjunct Professor for "Integrated Risk Analysis" at Stavanger University (Norway), Honorary Professor at the Technical University Munich, and Affiliate Professor for "Risk Governance" at Beijing Normal University. His research interests include risk governance (analysis perception, communication), stakeholder and public involvement in environmental decision making, transformation processes in economics, politics, and society, and sustainable development.

Frank Ulmer is a Senior Expert in Stakeholder Dialogues, Sustainable Development and Transformation at the nonprofit company DIALOGIK, a research institute for the investigation of communication and participation processes. He is the founder and director of a consultancy for strategy and dialogue, the Kommunikationsbüro Ulmer GmbH. His current focus of work is participation in climate protection policy making, transdisciplinary work on the Energiewende, and citizen's involvement in a sustainable municipal development. As a visiting lecturer at the Leadership Academy Baden-Wuerttemberg and the Verwaltungshochschule Ludwigsburg, he teaches participation, technology assessment, and agile administration. In general, his interest includes stakeholder and public involvement in decision making, transformation processes in economics, politics, administration and society, and education for sustainable development.

Anna Deckert works as a research associate at the nonprofit company DIALOGIK, a research institute for the investigation of communication and participation processes, and as a consultant at the Kommunikationsbüro Ulmer GmbH. Her research interests are the initiation and maintenance of citizen's involvement in a sustainable municipal development, a sustainable transition of the mobility sector, the role of guidelines for a shared participation culture and the added value of digital tools in offline participation. Her Master's degree in Sustainability Economics and Management (MA) at Carl-von-Ossietzky University of Oldenburg and research experience in living labs (Reallabor) provide her with a strong focus on and requisite know-how for transdisciplinary work.

About the editors

CHAPTER

1

Introduction

Ortwin Renn[1], *Frank Ulmer*[2] *and Anna Deckert*[2]
[1]Institute for Advanced Sustainability Studies (IASS), Berliner Strasse, Potsdam, Germany
[2]Dialogik Non-profit Institute for Communication and cooperation research, Stuttgart, Germany

OUTLINE

Traditional energy policy making is largely characterized by a technocratic vision in which the technical–natural sciences develop physically operational solutions for eminent problems, such as climate change or pollution from coal combustion. These blueprints of what works are then applied as a normative guidance for how society should design the appropriate policies to ensure their implementation. The technocratic model of energy supply is based on the ideal of objective knowledge about how to transform physical resources into energy services and how to make this transformation efficient, reliable and sustainable, and resilient. This concept of energy policy making is problematic for two reasons (Meyer, 2001: 165): First, the choice of what people need is already culturally influenced. The postulated assumptions about correlations between use of the physical resources, conversion technologies, and the utilization of energy services for human purposes are determined by preconceived cultural frames, which require their own explanation and questioning. Second the supposed chain of knowledge and management falls short. Knowledge alone does not change behavior. Although knowledge is a prerequisite for rational action, it does not replace the necessity of having normative priorities for weighing the consequences for the biophysical environment against other positive or negative impacts on the economy and society. At the same time, the balancing of these processes itself is dependent on social preferences. For example, the request to reduce the amount of fossil fuels for energy conversion can be met by a set of alternative actions: replace fossil fuels by renewable energy sources, make your energy system more efficient

The Role of Public Participation in Energy Transitions
DOI: https://doi.org/10.1016/B978-0-12-819515-4.00001-5

to produce more energy service from each unit of primary energy, or reduce consumption by demanding less energy services such as heat or power. Which of these three options or their combination will be implemented is not a question of technical feasibility not even economic optimization but rather of political preferences and public acceptance.

Energy policies need to be seen as being embedded in a socio-technical system in which technical, economic, political, and social factors interact (Renn, 2014). From a research perspective, studies on energy transformations need to include the consequences of human behavior on the choice of energy technologies and the everyday use of energy systems as well as the reflexive perception, assessment, and evaluation of human behavior when experiencing transformations of energy systems (Rip, 2006: 92f). Such an integrative approach includes five key objectives (see Dunlap et al., 1994; Becker et al., 2001; Renn, 2014):

- To gain systematic understanding of the processes of how citizens form preferences and positions about energy technologies and policies.
- To gain better knowledge of processes and procedures, that shape or enlighten the *social discourse(s)* about the right balance between different options of meeting energy service demand and, consequently, about ethically justifiable degrees of interventions into the natural environment on the basis of comprehensible and politically legitimate criteria.
- To investigate institutional processes and organizational structures that review, revise, and regulate individual and collective energy-related decisions.
- To identify political processes of how to include consumers and citizens in the decision-making processes about collectively binding energy policies.
- To investigate not only obstacles and barriers but also opportunities and incentives, which are associated with the inclusion of stakeholders and citizens not only in the design phase but also in the implementation of energy policies.

For this socio-technical perspective the need for participatory and inclusive policy making is crucial (Renn, 2008: 273ff; Scholz, 2011: 388ff). The definition of the energy problems, the question of which values to include, and the selection of methods used to consider and evaluate solutions all presuppose normative assumptions that need to be aligned to the preferences and values of those who will be affected. These requirements cannot be derived legitimately from the natural or technical sciences but result from social discourse. In this respect energy transformations rely on the involvement of affected persons and groups in the process of system selection and governance.

More than ever, discursive processes are at the center of a rational, integrative, and sustainable energy transformation. A discourse without a systematic knowledge base remains only superficial, and a discourse that hides the normative quality of the courses of action being considered violates the principles of democracy and individual freedom. In this respect all discursive processes must be measured by how they have conducted the integration of knowledge, values, and preferences.

The conditions and challenges for including stakeholders and the public at large for making energy system transformation successful differ among the European countries. This applies to the social and cultural conditions—for example, the acceptance of nuclear energy in France as opposed to Germany. But it also applies to the geographical conditions—for example, the abundant supply of hydropower in Norway as opposed to fewer hydropower options in other European countries. Coping with these regionally different challenges requires tailor-made solutions. However, all democratic and pluralist countries demand processes of legitimization based on public approval and at least some degree of codetermination (Naz and Leggewie, 2019). The main objective of this book is to first develop a systemic approach to participation and involvement and apply these generic principles and concepts to a set of case studies that all relate to the German energy transformation (Energiewende).

The challenges of implementing the energy system transformation in Germany are primarily characterized by a high per capita energy consumption; a strong dependency on fossil fuels on electricity, heat, and mobility; and the political framework conditions of a corporatist political culture. In concrete terms not only the discussions on the energy system transformation in Germany currently revolve around wind power, photovoltaic systems, and grid expansion, but also the level of consumption, successful system integration, and energy efficiency are major topics of the debate. This makes energy policy making rather complex and sophisticated.

The state of the energy system transformation in 2020 is sobering. Only about 15% of primary energy consumption in Germany is covered by renewable energies (Umweltbundesamt (UBA), 2019). At the same time, the demand for energy services in Germany continues to rise. This is probably the result of digitalization and a general rise in consumption levels in almost all areas of life (passenger numbers in air travel, car registrations, living space per capita, etc.).

Although Germany is one of the most advanced countries in expanding wind and solar energy, the speed of transformation is not sufficient to reduce the amount of fossil fuels significantly. The population is also divided: Many demand more rigid climate protection measures, including those that may change personal behavioral patterns (such as living without a car in urban areas), while others remain convinced that their personal lifestyle does not need to change.

This anthology assumes that in a plural and democratic society, public and stakeholder participation is an important if not crucial way of getting to broadly accepted energy policies and required changes in personal lifestyle. Participation in this book is regarded as an instrument that can lead both to increased acceptance of the local infrastructure and to new consumption patterns. However, it is not obvious of how to design effective and fair involvement processes and how to link scientific expertise with public values and preferences. This book will provide not only some theoretical insights but also many practical examples of how participation is designed and implemented in the present intensive phase of the German energy transition. It is well understood that the concrete participation projects described in this book cannot serve as recipes for similar projects in other countries and political cultures, but many conceptual approaches and procedures are certainly transferable to similar situations in other countries.

The basic message of this entire book is that major energy transitions toward more renewable energy sources and toward more energy conservation depend on the support of major stakeholders and the affected public(s). However, the methods and formats of how to include stakeholders and the affected public vary in the literature, and the empirical evidence is full of ambiguities. This book strives for both a clear theoretical foundation and sufficient empirical evidence for how generic concepts of participation can be used to design and evaluate public involvement and participation programs. This book also touches upon new institutional reforms of democratic decision-making as they appear as evolving under the light of public involvement. How should public input and public involvement be designed, structured, and organized so that it facilitates the transition toward a more sustainable energy future?

This book serves two main purposes: First it provides a conceptual approach to stakeholder and citizen involvement in a concrete transformation process, that is, replacing or reducing fossil fuel by increasing the efficiency of energy conversion, decrease of energy demand (sufficiency), and inclusion of renewable energy sources. These concepts form the basis for the empirical case studies that are described in the second part of this book, where concrete participation and involvement projects are outlined—predominantly from Germany since this country is one of the world's front-runners in planning to transform its energy system.

In Chapter 2, History of the energy transition in Germany: from the 1950s to 2019, Ortwin Renn and Jonathan Paul Marshall expound the emergence of energy policies in Germany, highlighting the major transitions over the last five decades. Following this introduction to the German context, Ortwin Renn und Pia-Johanna Schweizer introduce the six approaches to inclusive governance, their foundations, applications and lessons learned (Chapter 3: Inclusive governance for energy policy making: conceptual foundations, applications, and lessons learned). In Chapter 4, Energy transition and civic engagement, Radtke et al. illustrate how civic engagement in the context of the German Energiewende occurs in two rather opposite directions: self-organized groups, who run their own renewable energy plants and support the energy transition, and grassroots initiatives, who campaign against planned infrastructure for renewables. Subsequently Tom Morton et al. explore the social and political processes of leaving coal and their relationships to renewable transition in Chapter 5, From coal to renewables: changing socio-ecological relations of energy in India, Australia, and Germany. Andreas Klinke (Chapter 6: Cosmopolitan governance for sustainable global energy transformation: democratic, participatory-deliberative, and multilayered) broadens the perspective when describing an alternative approach to a cosmopolitan, democratic governance, which he believes is necessary to tackle the global challenge of energy transformation.

After exploring the theoretical and conceptual approaches to participation in the energy field several case studies of concrete participation or participation research form the remaining second part of this book. In Chapter 7, The Kopernikus Project ENavi: linking science, business, and civil society, Piet Sellke et al. describe the research project Energy Transition Navigation System (ENavi) that aims at delivering socially feasible, economically sound ethically just, and environmentally sound solutions for implementing the energy transition in Germany. In Chapter 8, Climate change policies designed by stakeholder and public participation, Jörg-Marco Hilpert and Oliver Scheel illustrate a case study of climate change policies designed by stakeholder and randomly selected juries in the State of Baden-Württemberg. In Chapter 9, Digital tools in stakeholder participation for the German energy transition. Can digital tools improve participation and its outcome? Anna Deckert et al. focus on the potential of visualization tools for making citizen participation more effective, particularly under complex decision conditions. This chapter describes and evaluates two case studies: the Forbach Pumped-storage Power Plant and the Traffic development in Herrenberg.

Citizen participation for wind energy in Germany and beyond is discussed in Chapter 10, Citizen participation for wind energy: experiences from Germany and beyond, by Gundula Hübner. Regina Schröter and Daniel Zirke analyze in Chapter 11, The contact group—public participation in the distribution network expansion in Baden-Württemberg, the contact group as method for public participation in the distribution network expansion in Southern Germany.

Daniela Setton argues in Chapter 12, Social sustainability: making energy transitions fair to the people, for an increase in participatory elements based on two major public surveys on the social sustainability of energy policy making. According to this chapter there is still a great need for participatory processes in order to make the energy system transformation in Germany socially sustainable. In the concluding Chapter 13, Conclusions, the three editors, Ortwin Renn, Frank Ulmer, and Anna Deckert, draw some general lessons from both the concepts and the case studies, in particular for practitioners and energy planners.

Although the focus of this book lies on Germany, the insights from the case studies as well as the theoretical foundations are valid for most countries with democratic governance structures and the explicit intention to reduce fossil fuel by renewable energy sources and new energy practices. The editors hope that this book will provide encouragement for more well-designed participation projects all over the world as a means to move step by step into a more sustainable energy future.

References

Becker, E., Jahn, T., Hummel, D., Stiess, I., Wehling, P., 2001. Sustainability: a cross-disciplinary concept for social-ecological transformations. In: Klein, J.T., Grossenbacher-Mansuy, W., Häberli, R., Bill, A., Scholz, R.W., Welti, M. (Eds.), Transdisciplinarity: Joint Problem Solving Among Science, Technology, and Society. Birkhäuser, pp. 147–152.

Dunlap, R.E., Lutzenhiser, L.A., Rosa, E.A., 1994. Understanding environmental problems: a sociological perspective. In: Bürgenmeier, B. (Ed.), Economy, Environment, and Technology: A Socio-Economic Approach. Sharpe, Armonk and New York, pp. 27–49.

Meyer, H., 2001. Quo vadis? Perspektiven der sozialwissenschaftlichen Umweltforschung. In: Müller-Rommel, F., Meyer, H. (Eds.), Studium der Umweltwissenschaften—Sozialwissenschaften. Springer, Heidelberg and Berlin, pp. 153–168.

Naz, P., Leggewie, K., 2019. No Representation Without Consultation. A Citizen Guide to Participatory Democracy. Between the Lines, Toronto.

Renn, O., 2008. Risk Governance: Coping With Uncertainty in a Complex World. Earthscan, London.

Renn, O., 2014. Towards a socio-ecological foundation for environmental risk research. In: Lockie, S., Sonnenfeld, D.A., Fisher, D.R. (Eds.), Routledge International Handbook of Social and Environmental Change. Routledge, London, pp. 207–220.

Rip, A., 2006. A co-evolutionary approach to reflexive governance—and its ironies. In: Voß, J.P., Bauknecht, D., Kamp, R. (Eds.), Reflexive Governance for Sustainable Development. Edward Elgar, Cheltenham, pp. 82–100.

Scholz, W.W., 2011. Environmental Literacy in Science and Society: From Knowledge to Decisions. Cambridge University Press, Cambridge, MA.

Umweltbundesamt (UBA), 2019: Primärenergieverbrauch. <https://www.umweltbundesamt.de/daten/energie/primaerenergieverbrauch#textpart-3>. (accessed 30.09.19.).

PART 1

Concepts of inclusive governance in the energy sector

CHAPTER

2

History of the energy transition in Germany: from the 1950s to 2019

Ortwin Renn[1] *and Jonathan Paul Marshall*[2]

[1]Institute for Advanced Sustainability Studies (IASS), Berliner Strasse, Potsdam, Germany

[2]University of Technology, Faculty of Arts and Social Sciences, Sydney, Australia

OUTLINE

The Role of Public Participation in Energy Transitions
DOI: https://doi.org/10.1016/B978-0-12-819515-4.00002-7

2.1 Authors: Ortwin Renn and Jonathan Paul Marshall

2.1.1 Introduction

The promises of an allegedly cheap, clean, and inexhaustible renewable energy supply contrasts strongly with the concerns about pollution and climate change arising from burning fossil fuels, and with the safety, weapons proliferation, and waste disposal issues arising from nuclear power (Martinez and Byrne, 1996). Yet implementing policy in favor of enhanced energy efficiency, renewable energy, and cutting greenhouse gas (GHG) emissions has not been straightforward in Germany. Major influences in the history of German energy policies have, on the nuclear side, included the incident at Three Mile Island in the United States, the Chernobyl accident in Ukraine, and most recently, the nuclear meltdown at Fukushima in Japan. On the coal side, disenchantment with coal started in the 1960s with concerns about the health effects of particulate pollution. In response, the technical elite emphasized the possibilities of making coal cleaner, and a high stack policy was introduced to dilute pollutants. However, high stacks were subsequently blamed for amplifying the "Waldsterben," or the slow dieback of German forests, through spreading nitrate emissions into the upper atmosphere and creating acid rain (Kenk and Fischer, 1988). Opposition to coal intensified with knowledge of climate change in the late 1980s, 1990s, and beyond.

In 1987, the West German government convened a meeting (Kohlerunde 1) with stakeholders close to the coal-mining and utility industry, which led to a program of reduction in coal mining but not to a complete phaseout. The participants concluded that securing the country's energy supply would require domestic hard coal. However, the negative impacts of mining and burning coal were being more and more acknowledged. Both the dominant narratives of coal as a solid domestic energy source and of nuclear as an inexpensive and environmentally friendly powerhouse began facing skepticism and open protest. This change in narratives triggered major changes and adjustments in public attitudes as well as in regulation and risk management (World Energy Council WEC, 2012).

This chapter gives a brief background and history of these policy complexities in the electricity market, in which coal and nuclear have made up more than 85% of Germany's electricity generation (in MWh). We describe the responses of regulators and policy-makers to contingent events, public opinion, economic pressures, and political commitments. Some abrupt policy changes resulted from actors taking unexpected strategic opportunities, as when antinuclear campaigners used the public outrage after Chernobyl and Fukushima to press forward new policy initiatives. Some resulted from unexpected consequences as with the high stacks producing dieback. Policy is always conducted in complex, uncertain, and unpredictable fields (both social and environmental) producing results that were neither foreseen nor intended by the dominant actors, undermining intentions, or leading in new directions. Policy-making with respect to a complex set of ecologies is not simply an a priori rational or predictable process, aiming to cater to divergent interests; it can also seem self-disrupting and paradoxical.

2.2 The European context of energy policies

Before discussing energy policies in Germany, it is important to understand the European context. Energy policy is at the heart of the European political and economic integration process (Matlary, 1997). After the war, one of the first steps toward the European Union (EU), initiated by the treaty of Paris, was the formation of the European Coal and Steel Community (ECSC or Montan-Union) in February 1953 (EU, n.d.). This was a conscious decision to make coal and steel part of the peace process (Schumann, 1950). Guisan (2011: 549-50) quotes Max Kohnstamm, the Dutch first Secretary-General of the ECSC, from an interview in 1999, as saying: "No genius was required to understand that we could not rebuild Europe without the Germans.... When I read the proposal for the ECSC I was struck as if by thunder: it was the way out of a vicious cycle, which was not only economic, but also ethical."

The process was complicated and somewhat messy, but making the European Coal and Steel Community advanced the integration process and boosted trade in both coal and steel. Writing in 1955, Friedmann claimed it was, "the first modern experiment in partial supranational government" (1955: 12). Friedman goes on to claim that the ECSC was heavily influenced by the American anti-trust philosophy, rather than the economic philosophy previously practiced by the signatories. This apparently came about because during the allied occupation of West Germany, there had been far-reaching "decartelization and deconcentration policies" and, so as not to discriminate against Germany, this idea was applied to the community as a whole (Friedmann, 1955: 5). However, this did not stop companies in the Ruhr from forming a cartel or the French industries from remaining nationalized (Friedmann, 1955: 21), but it set in motion the idea of an open market. Sanderson wrote (1958: 199): "It is hardly surprising that the experience of ECSC has been drawn upon in the drafting of the treaties for the European Economic Community and Euratom." Coal was central to the formation of the EU.

This strong influence of the European context is illustrated by four major policy developments which have affected Germany: (1) deregulation and liberalization of European energy markets, (2) climate change, (3) energy security, and (4) the introduction of the EU emissions trading system (EU ETS).

2.2.1 Deregulation and liberalization of energy markets in Europe

While the ECSC implied working toward open markets between states, this general move was intensified for energy supply in the mid-1980s, when a consensus on economic policies emerged among European business and policy elites, which assumed that "deregulated" and liberalized markets improved economic efficiency and general well-being. This conviction initially shaped policies in Great Britain, but was later adopted by all European leaders in the guise of EU internal market policies. While the effects of such policies are still contested (Buchan and Malcolm, 2016), they certainly increased the disruptive corporate influence on EU Energy Policy. In neoliberalism, protecting profit and securing the influence of established players become a prime aim of policy, and the energy industry aims at profit before attending to energy security, low pollution, low costs, or sustainability.

Previously, countries internal energy markets were relatively heavily regulated, with public monopolies dominating. Hence, the new paradigm added significant challenges for energy policy-makers while facing resistance from the old providers. Despite resistance, the markets were gradually liberalized and public monopolies challenged.

The principles of liberalization were not applied universally, or coherently, to all energy sources. This was particularly true for nuclear power, which was often offered protection from market competition through subsidies and other interventions, as its ability to raise capital from private investors was limited due to the uncertainty associated with large up-front investment costs, insurance costs, and uncertainty about licensing over its life span (Hatch, 1986).

2.2.2 Climate change—European, national, and international targets

Climate change has increasingly gained salience for policy-makers and in the public mind. Energy policy is a crucial means of influencing climate change, as the mix of energy sources determines emission levels of GHGs (in particular, CO_2) which generate such change. The climate change debate proved a major challenge to the acceptability of coal and fossil fuels generally, and thus challenged corporate profits in the liberalized energy market. Many players in the relevant corporate sector responded through ideological obfuscation, with market "necessities" often corrupting the flow of accurate information that confronted profit and, in turn, confusing policy processes (Beuermann and Jaeger, 1996).

Despite the confusion, measures to curb CO_2 emissions were undertaken on the international, European, and national levels. The 'Kyoto protocol to the United Nations Framework on Climate Change', adopted in 1997, called for GHG emissions to be reduced by 2008–2012 with EU countries accepting the obligation to reduce their GHG emissions by 8% from those of 1990. Some European countries, among them the United Kingdom, Germany, and France, established more ambitious national targets, others such as Poland and Czechoslovakia favored continuing coal use, foreshadowing future conflicts that have inhibited transition. In 2019, EU targets of zero CO_2 emissions by 2050 were said to have been blocked by the Eastern European states (Kuebler, 2019). Germany has decided to reduce fossil fuel use by an astonishing 80% by 2050 (Lauber and Jacobsson, 2015) and in 2018, in view of the Paris agreement on climate protection, Germany's environmental ministry announced the aim of approaching climate-neutral electricity production by 2050, which implies a total ban of fossil fuels in the electricity market.

Germany may, in turn, have had some negative effects on European-based emissions reduction. For example, when in 2013 after 5 years of negotiations, EU governments had agreed on emissions reductions from car exhausts, it was widely alleged that Angela Merkel moved to prevent this from passing, largely to benefit diesel production and German car manufacturers (Rueter, 2017; Monbiot, 2017). Germany has also supported biofuels, and the capacity of biofuels to reduce emissions is not obvious, but they can lead to destruction of forests and change of land-use from agriculture to fuel, thus potentially causing food shortages.

Various policy measures were established to achieve emissions targets, ranging from EU (and additional national) emission certificate trading schemes (see Section 2.4) to energy conservation incentives and climate change targets. The nuclear industry expected to benefit from these climate commitments, as it is virtually carbon-emission free after the plant is constructed. However, after the Fukushima accident, Austria, Switzerland, Italy, Denmark, and particularly Germany shifted their climate policies further toward renewable energy, with Germany deciding to phaseout nuclear energy altogether, whereas

France and the United Kingdom reasserted their commitment to nuclear power as a tool for reducing climate change.

2.2.3 Energy security—the revival of a well-known concern

Oil price hikes, unstable (or possibly hostile) political regimes in the Middle East and Russia, and depletion of North Sea oil and gas fields fed into growing concerns over the political and economic vulnerability of Europe to energy insecurity. These policy concerns seem to have diminished with the exploitation of more domestic oil and gas reserves through fracking technologies, despite people in many European countries (especially Germany) having strong environmental reservations against fracking and despite more than 60% of the fossil fuels consumed in Germany and more than 80% consumed in France still being imported (Schirrmeister, 2014). The imagined prospect of having more fossil reserves than previously assumed seems to have reduced fears over dependency. However, energy security remains a major topic of debate in Germany and other European nations (McCollum et al., 2014; Kleinert, 2011) and has strengthened the position of indigenous energy sources such as coal, nuclear power, and renewables.

2.2.4 The European emission trading system

In 2003, the members of the EU agreed to initiate an emissions trading system (EU ETS). The idea of the ETS is based on establishing a cap to all emissions for power production (Oberthür, 2019). The amount of emissions that fall below the cap is publicly auctioned giving a price for carbon emissions through the market, rather than through a stable carbon price decreed by the EU—another example of market liberalization in action. The EU compounded the complexities by deciding to hand out certificates for free to companies that already had licenses for using fossil fuels for electricity production or that used fossil fuels for industrial production (grandfathering). This effectively rewarded business for their previous pollution. Additional certificates were offered on the market for newcomers, but by taking advantage of increased efficiency in their power plants, the incumbents were able to sell their free certificates and earn an extra margin of profit. This has led to what has been called "a state of near-permanent political-economic crisis, with periodic market crashes" and questions "over the social and environmental integrity of carbon credits produced by offset projects that can be exchanged for European carbon allowances" (Bryant, 2016: 309).

As almost all certificates were given out for free, the price for each certificate dropped considerably, apparently saving various influential corporations a great deal of money, lowering corporate resistance to the project, and allowing even more emissions; thus, at least partly, undermining the point of the trading system. As a result, expectations that the ETS would reduce carbon dioxide emissions significantly were not met. After increases in demand for certificates in 2018 and 2019 due to an increase in economic recovery in many European countries, Europe witnessed a modest increase in the prices for ETS certificates, which may help facilitate the politically desired transformation from fossil fuels to renewable energy.

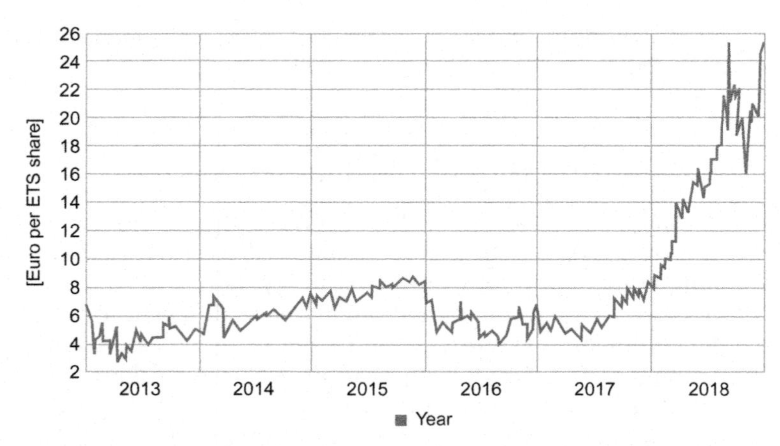

FIGURE 2.1 Overview of the price trend for ETS shares from 2014 to 2019. Source: *https://www.dehst.de/SPK/SharedDocs/downloads/EN/Auswertungsbericht_2017_Englische_Version.pdf?__blob = publicationFile&v = 2.*

Fig. 2.1 provides an overview of the price trend for ETS shares from 2014 to 2019. In summary, we can see the price began at about €6.5, fell to a low of about €3.0, slowly climbing from May 2017 until reaching a high of €29.46 in mid-2018, then falling back into the mid-2020s and reaching a new high of €30.0 in September of 2019 (not in the figure). Such a market volatility can allow additional rent-seeking. It is also possible that as coal-fired energy shuts down, the certificate cost of producing CO_2 will decrease, thus making pollution more economical. Furthermore, the system is weak as transport and heating account for about 40% of Germany's greenhouse gas emissions but are not yet covered by the EU's carbon market (Parkin and Wilkes, 2019b). The system also requires some independent body to verify the emission reduction from businesses, or businesses would have a financial incentive to underestimate emissions, so they could sell certificates to others. Eventually, it is hoped that a steady reduction of the number of certificates will lead to higher prices and reductions of emissions (https://markets.businessinsider.com/commodities/co2-european-emission-allowances).

2.3 The German national energy context

2.3.1 The energy situation in Germany

Germany is the largest energy market in Europe. Energy supply fell in the 1990s, partly due to restructuring the economy of the former German Democratic Republic (the so-called New Länder states), the decommissioning of some inefficient and polluting New Länder power stations, and the privatization of East German utilities and lignite mines, which initially led to large job losses and some levels of distrust in the new system (Boehmer-Christiansen, 1992). Between 2000 and 2018, the demand for electricity remained mainly constant and the demand for gasoline and diesel (for transport) increased, while the demand for primary energy sources for producing heat (low and high temperature) significantly decreased (AG Energiebilanzen, 2019). The latter is partially due to government subsidies for energy-efficient renovations in industry and private households, and the abandonment of coal burning for household heat.

Until 2010, half of all domestically produced energy came from coal and lignite; the rest came from nuclear, gas, and renewables. Since 2010, the share of renewables has dramatically increased while the share of nuclear has decreased (Renn and Dreyer, 2013; AG Energiebilanzen, 2019). Hard coal production and its use have declined, while lignite production has remained almost static over the last decade. The share of nuclear energy in domestic energy production, currently at 14%, is likely to reach zero by 2022 if the phase-out decision is not modified. Domestic gas production accounts for less than 10% in 2015 of the overall electricity production, while domestic oil production was negligible when compared with demand (AG Energiebilanzen, 2019).

Despite efforts to improve energy security through domestic production, energy imports still account for more than 60% of the total primary energy supply (Wettengel, 2019). This is particularly true for the heating and transport sectors, although less so for electricity. The share of imported energy per sector is illustrated in Fig. 2.2. In 2017, Germany already *imported* 51.2 million tons or 93% of the hard coal consumed. Its leading coal suppliers are Russia (38.5%), the United States (17.8%), and Colombia (12.7%). Hard coal amounted to 10% of Germany's primary *energy* use in 2018.

2.3.2 Energy industry

The German oil industry became fully privatized as refineries and distribution companies in the New Länder were handed over to private companies in 1992. The German oil market had a large number of players, although liberalization has coincided with major mergers in recent years (e.g., Exxon and Mobil, Total, PetroFina, and Elf, the merger of BP

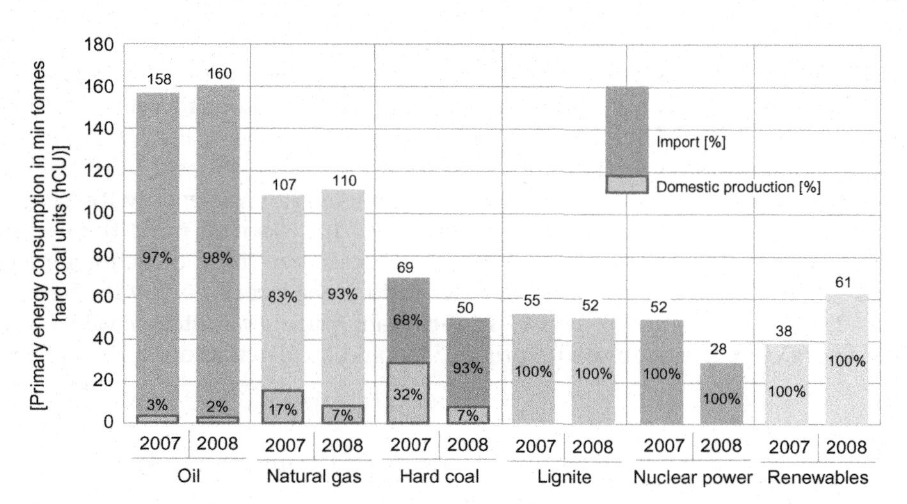

FIGURE 2.2 Import dependency by primary energy source 2007 and 2017 for Germany. Source: *https://www.cleanenergywire.org/sites/default/files/styles/gallery_image/public/paragraphs/images/german-energy-sources-import-dependency-2007-and-2017-1.png.*

and Veba Oil and Deutsche Shell and DEA Mineralöl), possibly demonstrating how market forces can produce oligopoly or less intense competition.

The German coal-producing sector is still the largest in Europe despite only one company producing hard coal in Germany, the RAG AG. This dominance was achieved when RAG took over Saarbergwerke AG (previously government-owned) and Preussag Anthrazit Gmbh (previously owned by Preussag AG). The RAG subsidiary, Deutsche Steinkohle AG, is responsible for RAG's German hard coal operations. The lignite-mining sector is dominated by Braunkohle AG, a subsidiary of one of the largest electricity producers, RWE. Again, it seems more monopolization than competition resulted from market liberalization. We will later discuss how market forces have boosted the use of high emissions lignite coal.

The German gas sector is also the largest in Europe, although heavily dependent on imports. In contrast to coal and oil, it is highly complex and competitive. It can be described as a multi-tiered, decentralized structure with a number of privately and municipally owned gas utilities. By 2002, there were 16 natural gas producers, 14 supra-regional companies six of which also imported gas (Ruhrgas, RWE, Wingas, BEB, EWE, and VNG), 15 regional distribution companies, and more than 700 local distributors and 11 gas dealers. This has not changed significantly over the last decade. Ownership of the companies was diverse from municipally owned local distributors to large privately owned supra-regional companies, with foreign companies holding significant shares (International Energy Agency, 2012: 18). In order to continue to increase energy consumption, Germany has made a deal with a Russian company (Gazprom) to build a pipeline to supply gas for fuel, despite protests from the Organization for Safety and Cooperation in Europe, and the governments of Poland, Ukraine, and the United States, that this renders German energy security subject to Russian imperial politics and that the Russian company Gazprom is notorious for its fugitive emissions (Roy, 2019; Auge, 2019). Using gas *may* reduce GHG emissions, but it does not lead to an end of them.

The German electricity sector is structured similarly to the gas sector. By 2014, there were 4 supra-regional companies, 56 regional utilities, more than 800 local utilities, and about 120 electricity dealers. Supra-regional companies generate electricity (80% of the market), transmit it over regional boundaries, and supply electricity to the final consumer. The four companies (RWE, EnBW, E.ON, and Vattenfall Europe AG) resulted from mergers: RWE acquired VEW Energie AG in 2000, VIAG and VIBA merged in 2000 to form E.ON, and the Swedish company, Vattenfall, acquired majority ownership in VEAG, HEW, and BEWAG. However, with the Energiewende, the number of prosumers (generating solar energy for personal consumption and selling the surplus to the national grid) has risen to more than 40,000, thus changing the energy supply sector dramatically. About one-third of self-generated solar power is fed into the national grid, while another third is self-consumed by the home or business (Meza, 2013). This change primarily results from price guarantees based on the EEG (Renewable Energies Act, inaugurated April 1, 2000) and local action not liberalization.

2.4 Electric power regulation and policy-making

2.4.1 Overview

Germany's political system is shaped by federalism. Consequently, Federal and state governments are involved in policy-making. All energy legislation is planned and adopted at Federal level, but state-level governments are responsible for implementing federal laws (Hatch, 1986). In addition, state-level governments can develop their own programs such as promoting renewables and competition. This division can add complexity to policy dilemmas, with potentially competing interests, but also means that decisions cannot be put through entirely unopposed if they are inadequate.

On the Federal level, the main responsibility for energy policy and research and development lies with the Federal Ministry of Economics and Technology (BMWi). The Federal Ministry of Environment, Nature Conservation and Nuclear Safety (BMU) is in charge of environmental policies, including climate change, nuclear safety, and the impacts of fossil fuel combustion. Promotion of energy conservation and efficiency is undertaken by the German Energy Agency (DENA), established in 2000 by the government and the German reconstruction bank, Kreditanstalt für Wiederaufbau (KfW) in order to bring together the various players in the energy sector, help in implementing energy efficiency policy, and promote renewable energy sources, climate change mitigation, and sustainable development. The regulation of competition in energy and electricity markets is the responsibility of the Federal Cartel Office (FCO) and, in some more local and regional cases, the state-level competition offices. In addition, special committees have been set up such as the Committee for Sustainable Development in 2001 and the Council of Sustainable Development (comprising representatives of all relevant interest groups in the area of energy and climate change policy). The latter developed a National Strategy for Sustainable Development that was adopted by the government in April 2002. Furthermore, energy policy is often guided by reports and discussions conducted by independent expert panels and institutes. So, there are a relatively large variety of inputs into policy, which again adds both complexity, variety, and awareness of alternatives.

With respect to the coal-mining and coal-using industries, the German government pursued a double strategy. Partly due to the large difference in prices between imported and domestic coal and partly fueled by the climate protection movement, the German government decided as early as 1987 (Kohlerunde 1) that they would curb coal production (at that time only hard coal due to the separation of East and West Germany) and later in 2005 (Kohlerunde 2) phaseout the use of hard coal in Germany by 2020. During this period, hard coal was phased out, and the cheaper and more emissions intensive lignite became dominant in coal-fired electricity production due to market economics. This produced the paradox that the mechanisms of the Energiewende withdrawal from coal may have increased emissions. In 2018, the Coal Commission or "Commission on Growth, Structural Change and Employment" was established by the Federal Government in conjunction with the States in which lignite coal is produced, to phaseout lignite coal and to transform Germany's energy system into a climate-neutral condition. It made the recommendation to phaseout all coal mining and prohibit utilities from using coal for electricity production by 2038. At the same time, the government launched special social programs to ensure socially and regionally acceptable phaseouts in order to preserve the Energiewende's acceptability to voters in coal-rich areas, especially in the East (Reizenstein and Kapop, 2019). The plan has been, and still is, to reduce unemployment resulting from the closure of the mines and to give financial aid to those who would not be able to continue working.

One of the most important open discussion platforms shaping the energy policy of the previous government was the Energy Dialogue 2000, led by Federal Ministry of Economics and Technology and the Forum for Future Energies, with the participation of political parties, state-level governments, companies, organized labor, and environmental organizations. This platform recommended the phaseout of nuclear energy and the slow termination of fossil fuels.

After Fukushima, the ad hoc Commission on the Future of Energy Policies (the so-called Ethics Committee) provided the main direction and justification for both the accelerated phaseout of nuclear energy and the roadmap for an energy supply system dominated by renewable energy sources (Ethics Commission, 2011). Even the nuclear industry did not openly oppose the recommendation of the ethics committee, although they challenged the "ambitious" time plan and asked for help sustaining nuclear expertise in dealing with continuous tasks such as waste processing and depositing, decommissioning of nuclear power plants, and participation in international safety and security research. Both nuclear phaseout and reduction of fossil fuels were the two major components of the German Energiewende. The third element was the dramatic increase in projected energy efficiency. Recommendations by the ethics committee included a phaseout of coal from the power market by the end of the century with an 80% reduction by 2050. All three recommendations were formally approved by the German government in the fall of 2011 and even sharpened in the aftermath of the Paris agreement on climate protection.

2.5 History of energy policy development

2.5.1 Energy policy before 1986

West Germany's energy policy between 1945 and 1973 was, within the framework of the ECSC, characterized by a relatively low degree of state interventionism and a broad acceptance of both nuclear and coal as energy sources. In the first decades after the war, West German energy policy was essentially coal policy due to the economic and political power of the coal industry (Hatch, 1986). East Germany, especially in Lausitz, would become the world's largest producer of lignite and famed for its air pollution and destruction of the countryside (Morton and Müller, 2016: 282). In 1950, coal accounted for almost 90% of West Germany's primary energy consumption for electricity and heat (excluding transportation). More than half a million people were employed in the coal industry throughout the 1950s. Most of the industry was located in the Ruhr basin in North-Rhine-Westphalia, one of the economically and politically most powerful of the eleven Länder in the Federal Republic. Coal-mining workers were organized in a powerful and influential trade union, I.G. Bergbau und Chemie, which was closely affiliated with the Social Democratic Party. Coal was an unchallenged champion.

Nuclear energy started to become important in the late 1950s. Pro-nuclear interests found a partner in the specially established Ministry of Atomic Affairs (founded 1956) which, under the social-liberal coalition, was abandoned and integrated in the Ministry of Research and Technology in 1962. In 1967, the first commercial orders were placed for two light-water reactors (LWRs) by German utilities. Two years later, three more were ordered, and in 1971, five more orders were placed. By 1973, all 10 projects had permits to begin construction, with additional plants planned for the near future. Two companies, Siemens and AEG, with licenses for LWRs by Westinghouse and GE, respectively, took the leadership in the plant construction industry, although AEG gave up nuclear power construction in the 1970s. The companies were encouraged by the government to establish the joint reactor development company Kraftwerk Union (KWU). When AEG discontinued its nuclear ambitions, KWU became a sole subsidiary of Siemens.

In addition, Germany supported the development of a Thorium high-temperature reactor (THTR), but support ended in the early 1980s as no commercial company was willing to build them and the only commercial test reactor in Hamm-Uentrop experienced significant increases in construction costs and major breakdowns and failures during its short operational phase from 1885 to 1989. The government was also involved in developing a reprocessing plant to close the nuclear fuel cycle. Basically, despite governmental support, nuclear energy experienced uncertainty even in the early days and, as we shall see, there was considerable opposition to it.

The cutbacks in oil supply by members of Organization of Arab Petroleum Exporting Countries, in support of opposition to Israel, which led to the oil crisis of 1973/1974, fundamentally transformed energy policy in Germany from relatively limited interventions to protect coal companies and their employees, into a comprehensive policy aimed at securing low-cost energy supply, while respecting environmental protection. This shift was reflected in the 1973 Federal energy program aiming at increased energy independence and security. Some years later, energy conservation was given formal priority with the second revision of the energy program in 1977. Despite disputes over the provision of funding (between state level and Federal level), as well as the type of measures (regulation vs market incentives), a major package was agreed in 1979 and confirmed in a third revision of the energy program in 1981. By the time of this revision, coal was seen as a domestic resource and seven new coal-fired power stations were commissioned. In addition, almost a dozen nuclear power plants were ordered, with others planned for construction later. In spite of major debates on nuclear safety, most elites in Germany supported the expansion of coal and nuclear as the best guarantees for energy security.

Simultaneously, in the 1970s, the environmental movement and public protest against nuclear power became a strong political force, leading to the establishment of a new party, the Greens (Renn, 2008, p. 60). The environmental movement also targeted coal-fired power stations and open-pit mining but protests over these issues remained at a low intensity to avoid fighting on two fronts concurrently. The anti-coal coalition of the early 1970s was largely confined to the local level.

Major antinuclear power protests started in 1975 with ad hoc citizens' initiatives (the so-called Bürgerinitiativen) in the state Baden-Wurttemberg where a large plant (Wyhl) was approved for construction. The initiatives gathered signatures of local citizens objecting to the construction of the plant and appealed to the administrative court. Arguments in these local campaigns initially revolved around local issues (Hatch, 1986). Later, all sites for planned nuclear power plants became the targets of major protest movements and demonstrations. Public outrage was specifically strong at the site of Brokdorf where demonstrators and police engaged in fierce battles.

The increasing relevance of nuclear energy and the increasing public opposition triggered a response by government, the so-called "Bürgerdialog Kernenergie," which aimed at regaining trust by promoting a propagandized consensus on nuclear energy. The engagement of the Federal government with the public took several forms (newspaper advertisements, information brochures, and public discussions) and changed its content over time in response to the broad concerns of the public, from purely safety-related issues at the beginning to less contentious issues such as the desirability of technological progress. However, the project failed, polarization remained, and violent clashes between protesters and police propelled the debate into a national showdown between government and protesters. The communication strategy may even have unintendedly brought the debate to a wider public.

Despite considerable success in mobilizing the public and putting nuclear energy on the national political agenda, the crucial battleground for German protesters were the courts. The courts dealt with nuclear issues from a wide range of perspectives not just energy policy, expanding the government's problems. While, for example, a construction stop for the plant in Wyhl was issued on the basis of safety concerns, the decision to stop construction of Brokdorf plant in the state of Schleswig-Holstein arose because of the unresolved problems of storage and disposal of radioactive waste. An administrative court in Lower Saxony accepted the arguments of a pharmaceutical company, which feared contamination of their medical products because of their close proximity to the nuclear reactor. In another significant case, the prerogative of the state executive to grant a license for the construction of the FBR at Kalkar was questioned by the court in the light of security and waste storage issues.

By the end of 1977, work on 3 of the 13 plants under construction was halted. In response, the industry pointed to the economic costs of reversal in terms of employment and technology export opportunities, two key issues in economic debates in Western Germany and two key factors justifying liberalization (and, in turn, being justified by giving the market priority), even if nuclear energy needed state subsidies.

In spite of fierce demonstrations and the court cases that prevented the construction, or the operation, of several nuclear plants, the protest movement had little success in halting pro-nuclear polices or influencing the majority of German people to develop an antinuclear attitude (Wagner, 1994). Until 1986, all parties except the Green party supported nuclear expansion. The early 1980s even witnessed a gradual improvement of the prospects for nuclear power, despite the ongoing political successes of protestors. In particular, the second oil crisis (with its skyrocketing prices) made it easier for the SPD-FDP government to renew its commitment to nuclear power in the third revision of the energy program in autumn 1981. Federal and state-level governments collaborated to speed up the licensing processes so that in February 1982, after a delay of 5 years, the construction of three new power plants was authorized by the Interior Ministry in "convoy." The three permits were granted on the basis of an approved standardized design and a governmental assessment that the criteria for waste disposal were satisfied.

Under the shadow of the controversies about nuclear energy, the environmental debate on coal and its impacts had little resonance (Lauber and Jacobsson, 2015). In West Germany, new open-pit mines for lignite coal were licensed in the mid-1980s despite local protests. During the first coal negotiations in 1987 (1. Kohlerunde), the need for reducing coal mining in Germany was acknowledged but the argument for national energy security dominated the discourse. Hard coal production was supposed to steadily decrease, but still contribute a major share to the domestic electricity production. Coal was regarded as a domestic energy source that both guaranteed energy security and stability on the labor market (Brauers et al., 2018). The climate change debate had only just started, and the connection of climate protection with Germany's coal program had arrived on the agenda of most environmental NGOs. They were, however, reluctant to fight against both nuclear and coal. So, nuclear and fossil fuels coexisted until 1986. This was the year when the nuclear industry experienced its greatest blow: Chernobyl.

2.5.2 Energy policies between 1986 and 2011

The Chernobyl accident had a major impact on public opinion and triggered substantial policy changes by the federal government. The most obvious change was the establishment of the BMU, the federal ministry for environment, nature conservation, and reactor safety. Before 1986, environmental issues were handled by the ministry of the interior. One of the first ministers appointed to head the BMU was Klaus Töpfer, who later became the director of UNEP, and after Fukushima was asked by Chancellor Merkel to chair the Ethics Committee on the Future of Energy Policies in Germany.

After Chernobyl, the social democrats (SPD) adopted more critical positions toward nuclear energy, while emphasizing pro-coal policies as part of their energy security program. At that time, the antinuclear movement was not just powered by urban radicals but also by members of the environmentally conscious middle class and conservative farmers objecting to state-imposed industrialization of the landscape (Morris and Jungjohann, 2016: 4, 17ff.). Coal had fairly solid support as the workers in Germany's coal mines were mostly organized in unions and were traditionally Social Democratic Party voters. Similarly, the conservative Christian Democratic Party was influenced by the coal lobby (the steel industry, local coal communities, and the unions) to sustain support for a national coal strategy. This strategy had two components. First, coal subsidies for uneconomic hard coal mines were guaranteed in a National phaseout plan until 2018 (Kohlerunde 2). The hard coal industry could continue operations until this date but was obliged to phaseout mining operations. By the end of 2018, all hard coal mines had been closed. Subsidies for hard coal mining amounted to 327 billion Euros from the Federal and State Government between 1970 and 2014 (http://www.foes.de/themen/energie/stein-und-braunkohle/).

For economic and energy security reasons, lignite was not included in the phaseout plan. Lignite was regarded as a national reserve that should continue to be supported, because it was cheaper to mine than hard coal and still a domestic resource, even if more polluting. This only changed after the events in Fukushima and the drafting of a national energy transition plan (see Section 5.3). Struggles continue over the expansion of lignite mines up to the present day (2019).

In the time period from 1986 to 1998 (during the Kohl administration), nuclear energy was severely attacked by the opposition parties. Even before 1998 when the SPD took office on the federal level, nuclear policies were strongly affected by the growing opposition from the SPD. The conflict between the governing conservative party (CDU) and the SPD led to the collapse of collaboration between states and the federal government on the issuing of nuclear licenses. SPD-led states actively started to implement legislation consistent with the official SPD policy of exiting nuclear power (Kern and Löffelsend, 2003). These state-level attempts were bound to fail because the conservative and the liberal party with command over the federal government could sustain their pro-nuclear position on the federal level until 1998. Once facilities (plants and storage sites) were up and running, it proved difficult for state-level governments to shut them down. As long as plant operation was "orderly," the Atomic Law protected them from closure. However, the new levels of conflict and inconsistencies between the rhetoric of the pro-nuclear advocates and the reality of nuclear power production, with many safety breaches and critical incidents, emphasized that nuclear energy was vulnerable.

Environmental concerns were also raised about lignite coal, which is far more polluting than hard coal. In addition to more extensive destruction of land and agriculture, the average CO_2 emission per Kilowatt-hour amounts to 0.41 kg for lignite coal in Germany and to 0.34 kg for hard coal (Volker-Quaschning, 2015). Lignite is also heavier for other emissions, in particular nitrogen oxides. However, lignite was much more profitable as it could be produced in cheaper open-pit mines, and this is important given market liberalization. The lignite industry was able to expand its operation in western North-Rhine-Westphalia. New pit mines such as Hambach were initiated in the late 1970s and villagers were relocated elsewhere in the early 1980s to provide room for more mining activities. As already mentioned, in Eastern Germany, there was a strong tradition of lignite coal mining in the Lausitz area, which basically depended economically on mining activities. There was a local protest but little national attention as most environmental groups were still concentrating on fighting nuclear energy. Furthermore, the phaseout of hard coal was already highly controversial and the consensus emerged that Germany would at least need its lignite coal reserves to secure energy supply.

Unification in 1989 had a significant effect on the perception of coal. Germans were suddenly confronted with partially contaminated or devastated areas in the Eastern coal-mining districts (often close to areas with a high concentration of polluting chemical factories). Many landscapes looked like "moon structures." While these impressions confirmed the image of coal as dirty, they also encouraged many pro-coal activists, union members, and politicians to demonstrate that "re-naturalization" efforts could compensate for the destruction in both "East" and "West." By re-naturalization, the West could show the East, the benefits of capitalism. Re-naturalized landscapes, such as the area near Bonn in North-Rhine-Westphalia (Ville), became popular tourist sites with newly grown forests and artificial lakes. Similar efforts were undertaken in East Germany and more than 4 billion Euros were spent turning devastated landscapes into parks or recreational areas (not all very successfully). The intended message was clear: coal has its negative impacts, but with modern technology, emission control, and landscape re-engineering, one could master these impacts. The debate was less about whether coal mining and coal-fired power production should be continued but rather in what form, and to what degree, and how the deleterious effects on the countryside could be repaired.

However, there was some degree of friction between East and West. As one author said at the time: "The difficult process of 'integration'... has made people in the 'two' Germanies more aware of their separateness than ever before. Many of their interests no longer coincide. Many people in the east perceive current events as a self-interested takeover by West German (and other) enterprises against which they as yet have few defences" (Boehmer-Christiansen, 1992: 197).

There were perhaps more difficulties in the re-unification process than had been expected, and the transition, change of currency, and privatization of industry (generally leading to collapse and Western corporate takeover) were rapid, which may have led to frictions between federal and state governance which still affect energy policy today and produce expectation of political resistance in the East.

During the 1990s, climate change became more salient to the long-standing background debate on the negative impacts of mining and burning coal. However, the main problem for the traditional hard coal mining industry was not environmental but economic. German coal was much more expensive than imports from Poland or Australia. The growing environmental concerns of NGOs and environmental activists added to a slow process of financial retreat from coal but it was not the major motive for the industry to reduce coal-mining activities in the Ruhr and the Saarbrücken districts, the two main areas in Germany with a tradition of hard coal mining. Indeed, the engagement of social democrats and unions kept many coal mines alive to sustain employment (Lauber and Jacobsson, 2015). The two major coal agreements (Kohlerunde 1 in 1987 and Kohlerunde 2 in 2015) confirmed both the need for reducing the amount of coal for energy generation, and the national interest in having a substantive national reserve of coal, in this case lignite, for reasons of energy independence. It was hard to reduce the contradictory policy tendencies centered around issues of climate change, emissions reduction, lowering coal consumption, and retaining energy growth and security.

A coalition of the SPD and Green Party won federal power in 1998 and things began to change. After its inauguration, the new government emphasized sustainable development in its energy policy. Its three key principles were as follows: (1) supply security, (2) economic efficiency, and (3) environmental compatibility. The government identified several key areas of action: (1) the mitigation of climate change, (2) promoting energy efficiency, (3) continued use of domestic coal and lignite, (4) creating more competition in the liberalized energy markets, (5) promoting renewable energies, and (6) creating a level-playing field for energy companies throughout Europe. This program was complemented by the phasing-out of nuclear energy (International Energy Agency IEA, 2007).

The incorporation of energy security into the catalog of main objectives recognized the relatively high dependence of Germany on energy imports and the ongoing depletion of local fossil fuel reserves. The hope was that further market liberalization would ensure efficient usage of energy, diversify supply sources, and improve energy services and energy trade. The government entrusted decision-making on investment, new capacity, and timing, to the energy industry, to an extent surrendering its power, even while attempting to change energy policy. After 2011, this changed when the government launched the "Energiewende." The energy industry is now highly regulated in Germany and has converted into a sophisticated governance structure involving many players: government, companies, NGOs, and citizen initiatives.

Initially, environmental compatibility was to be achieved through a mixture of voluntary agreements with the industry, e.g., the 2000 "Agreement on Global Warming Prevention" (which was meant to reduce CO_2 emissions per unit of output by 28% by 2005 and overall GHG emissions by 35% by 2005, when compared with 1998 levels), the eco-tax, the promotion of renewables (through the Renewable Energies Act, the Market Incentives Program and the 100,000 Rooftops Solar Electricity Program), and energy efficiency improvements. Overall, the policy aimed at a continuous phaseout of nuclear energy, a reduction of fossil energy production and consumption, an increase of energy conservation, and an increase in renewable energy production, with the goal of curbing CO_2 emissions by more than 20% by 2020 and more than 40% by 2030.

Both parties agreed that phasing out nuclear power was a top priority. However, the exact method of achievement invoked controversy between the coalition partners and within the Green Party itself. The SPD wanted to avoid any compensation claims by the utilities, hence, favored a phasing-out in consensus with the industry. The Green Party was split. The more pragmatic wing, primarily those who had been working in governments at the state level, particularly in the Ministry of Environment in Hesse, knew about the legal and political difficulties of imposing a phaseout on electricity utilities. Hence, they were more sensitive to the problems of demanding an immediate shutdown of plants, an end to the shipment of nuclear waste by trucks (loaded with so-called Castor casks in which spent nuclear fuel was stored) and the export of nuclear waste. As a result, they sought to explore the opportunity of a constitutionally sound, and compensation excluding, law that would lead to a gradual phaseout of nuclear energy. The radical wing of the party, closely affiliated with antinuclear activists, wanted an immediate exit.

After a long debate, an agreement was reached in 2002 with the limits of operation, in general, fixed at 32 years. The amount of electricity produced by each reactor was fixed (favorably for the utilities), and these amounts could be transferred from one reactor to another, thereby extending the lifetime of individual reactors beyond 32 years.

After a conservative victory in 2005, the decision to phaseout nuclear energy was again up for debate and change. Industry and utility companies took advantage of a "sympathetic" government to urge a change in the timing of the phaseout and using nuclear energy as a long-term "bridge" between the fossil and renewable eras of energy production. After its re-election in 2008, the conservative government proposed a law extending the time limits for the phaseout and providing even more flexibility for energy utilities (Renn and Dreyer, 2013). In 2010, Chancellor Merkel was about to complete an agreement with the large power companies, which would both enable a raise in the spent fuel tax and postpone the phaseout of nuclear energy. Although the government denied any link between the two measures, the German public and most of the media were convinced that this was a deal made in secret. Opposition parties charged the government with compromising public safety in exchange for increased revenues. Many environmental groups organized large demonstrations against the government's plans (Buchholz, 2011).

2.5.3 Energy policies between 2011 and 2019

In the middle of this heated debate, the Fukushima accident occurred and this contingent event altered the political balance. A few days after the accident, the conservative party lost an election in the key German state of Baden-Württemberg, and, for the first time in German history, the Green party won the majority of the votes and could make one of its leading party heads the premier of the state.

The federal government responded by deciding to shut down seven of its oldest nuclear units and not re-open one that was out of operation at that time (Renn and Dreyer, 2013). In addition, they requested the German nuclear safety commission to conduct a stress test on all remaining 11 nuclear units in Germany (DG Energy, 2011). Within a few days of the accident, the government also established the so-called "ethics committee" on the future of energy policies in Germany. This committee did not include experts on energy and nuclear risks. It was composed of elderly statesmen from all political parties, functionaries of the major scientific organizations in Germany, social scientists and philosophers of ethics, and, as usual in the corporatist regulatory style of Germany, representatives from the major civil organizations such as employers' associations, unions, and the two major religious groups—Catholics and Protestants. (One of the authors, Ortwin Renn, was a member of this committee.) Their mandate was to develop a roadmap for designing energy policies for the future (Ethics Commission, 2011). The ethics committee had only 6 weeks to come up with its recommendations since the shutdown of the older reactors needed to be legally confirmed within that period if compensation payments were to be avoided, another contingent factor that may have affected what was considered. After 6 weeks, the ethics committee recommended the phaseout of nuclear energy within 10 years while promoting energy efficiency and installation of renewable energy sources. They recommended the reduction of the amount of CO_2 emitted in 2015 to 20% of the emission levels of 2005. They also recommended that the government should establish both an auditing committee to make sure that the Energiewende would run smoothly, and an energy public forum to boost acceptance of the new energy policies (Ethics Commission, 2011; Renn and Dreyer, 2013).

At the same time, the Reactor Safety Commission issued its report on nuclear safety in Germany. They did not detect any major weakness in German reactors and thought there was a high degree of resilience even for events beyond the design parameters used for licensing these reactors. However, they also concluded that older reactors would be vulnerable to large earthquakes and all reactors to terrorist attacks (Bruhns and Keilhacker, 2011). In June 2011, the German Parliament adopted the recommendations of the ethics committee. All parties in the German Bundestag voted in favor of the Energiewende, with only a few representatives voting against the law or abstaining. The law mandated a phaseout of all remaining nuclear power plants by 2022. In addition, the new law included provisions to reduce the share of fossil fuel from over 80% in 2011 to 20% in 2050. Energy efficiency was to be increased by 40% when compared with the average efficiency rates of 1990. The reactors shut down immediately after the accident, remained closed.

In a poll taken directly after the Fukushima accident, 78% of the German public approved the new stance of the German government (Renn and Dreyer, 2013). The IPSOS international survey, conducted in June 2011, showed that only 9% of the German population believed that nuclear energy could provide a viable long-term solution to energy needs. This was the lowest percentage of the 24 countries in the survey (Germany shared that position with Sweden). The strong antinuclear attitude among the German population, in conjunction with a strong citizen protest movement against nuclear energy, almost certainly reinforced political party support for the transition toward a non-nuclear future in Germany, and the rapid transition of conservative forces to a new outlook.

Between June 2011 and 2015, the new Energiewende law had major effects on energy supply and consumption. The share of renewable energy in energy production increased dramatically from 17% to 28% and the share of nuclear fell from 23% to 16%. However, the share of lignite increased but stabilized after 2014. This eventuated due to favorable weather, with much sun and wind, with the spot market prices for electricity falling to an all-time low, which led to a situation where gas or even hard coal–fired power plants were not competitive. Only power plants running on cheap (CO_2 intense) lignite were able to make a profit. As a result, the share of lignite coal increased at the same rate that the nuclear share decreased.

A paradoxical situation emerged: the more that Germany invested in the Energiewende (it poured in more than 27 billion Euros in energy subsidies, financed by energy consumers), the more the amount of CO_2 increased because lignite remained competitive in a liberalized energy market (Chrischilles and Bardt, 2016). The government tried hard to overcome this problem by suggesting limits for the use of lignite or offering new payment schemes with flat rates for energy security (which would keep gas-fired stations for backup service in business). At the end of 2019, the new agreement on phasing-out lignite coal until 2038 was formally adopted by the German government (still pending confirmation in parliament).

In the years between 2011 and 2015, lignite coal production in Germany averaged 180 million tons per year, largely due to the phaseout of hard coal. In 2015, the share of lignite in electricity production dropped for the first time after 2011, though by a mere 0.3% when compared with the 2014 level. For the first time, also, renewable energy sources provided more energy in terms of Kilowatt hours (40.9%) than lignite (39.4%). While some of the local population at the lignite sites in East and West Germany were encouraged by the new boom of lignite and lobbied hard for a program that kept lignite as the main fossil fuel source in Germany, other locals protested against expansion of lignite mines and the destruction of their self-sufficient, renewably powered, villages (see Morton and Müller, 2016). Despite studies showing that Germany had enough mines for the foreseeable future, villages where still being reported as going to be abandoned for the expansion of lignite mines in 2019 and prompting continued protests (Schauenberg, 2019).

Fig. 2.3 provides an overview of the development of energy sources for electricity production from 1990 to 2018 (adapted from AG Energiebilanzen, 2019). Most dramatic is the increase of renewable energy sources from 20 TWh in 1992 to 226 TWh in 2018. The phase-out of nuclear energy has already resulted in a decrease of nuclear share of the power markets from ca 160 TWh in 1990 to 74 TWh in 2018. The share of hard coal decreased from 150 TWh (1992) to 83 TWh (2018). While hard coal mining is phased out in Germany, the use of exported coal in power plants is not. Lignite coal experienced only a slight decrease from around 165 TWh in 1990 to 146 TWh in 2018.

Although renewable energy sources supply more than half of the installed capacity in Germany, only between 35% and 40% of all electricity is produced by them (https://www.ise.fraunhofer.de/content/dam/ise/en/documents/News/Stromerzeugung_2018_2_en.pdf). In total, the renewable energy sources solar, wind, water, and biomass produced ~219 TWh in 2018. This is 4.3% above the previous year's level of 210 TWh. The share of public net electricity generation, i.e., the electricity mix that actually comes from the socket, was over 40%. The share of the total gross electricity generation including the power plants of the

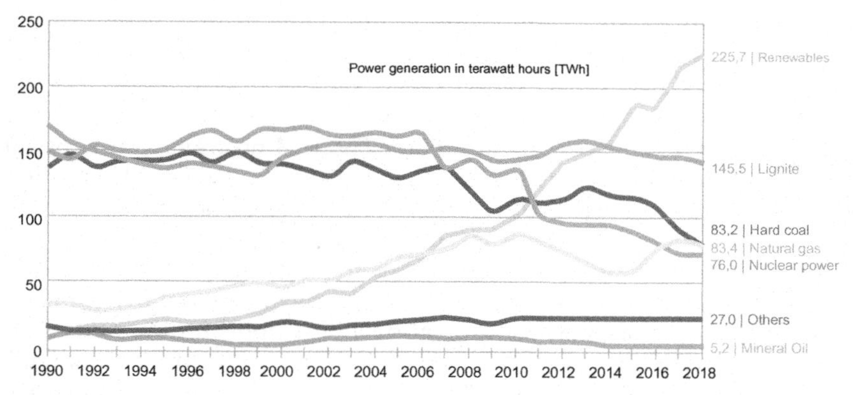

FIGURE 2.3 Gross power production in Germany 1990–2018 by source. Source: *Adapted from AG Energiebilanzen 2019.*

"companies in the manufacturing industry as well as in mining and quarrying" is ~35% according to the Statistics of the Association of German Utilities (BDEW; https://www.bdew.de).

This gap between installed capacity and net production of electricity is due to the volatility of the energy supply (BMWE, 2016). If the wind does not blow or the sun is not shining, the installed capacity produces smaller amounts of electricity than otherwise, and this requires more backup capacity storage or flexible smart grid structures than are currently available (Kleinert, 2011; Goutte and Vassilopoulos, 2019). In addition, public opposition to the extension of high-voltage power grids has led to the wastage of electricity from off-shore wind installations, since the power generated there could not be transported to the south of Germany where most of the demand is located. It is possible these ongoing politico/technical problems indicate a significant gap in moving from goals of the energy policy to an overall plan by which the goals may be attained. There is also a major political problem of getting public support for the necessary changes to transform the energy systems from a fossil-based, *demand-driven* electricity system to a renewable-based, *supply-driven* system (see section 6).

Levels of incoherence and dispute remain. In late 2018, Greenpeace and other environmentalists suggested that Germany was allying with Central European states to obstruct plans to raise the EU's 2030 nationally determined contribution climate targets under the Paris Agreement, to 40% of 1990 levels, and was "refusing to clarify its stance on proposals by other EU states to phaseout subsidies for coal power plants and force them to meet tougher pollution rules." The German energy minister continued to support the further expansion of Lignite mines (Boren, 2018; Energywire, 2019). Germany was widely expected to fail its 2020 emissions targets of a 40% cut from 1990 levels, possibly because lignite burning was not declining, or because of increased emissions from vehicles and buildings (Rueter, 2017; BMU, 2019). This is despite the base year of 1990 meaning that very emissions heavy lignite coal power stations were still operating in East Germany. Hans Josef Fell, the initiator of Germany's 2000 renewable energy feed-in tariff legislation, has been reported as saying "The so-called coal commission is counter-productive since it delays reduction of private coal assets," and "Climate protection means reducing GHG emission to zero, implying an immediate halt of fossil fuels." Fell further wondered why the German government was so concerned about the loss of coal miner's jobs when they had apparently been unconcerned "when some 80,000 jobs were lost in the solar industry over the last [few] years" and complained that "There is no transition happening in the automotive and building sectors" (Stam 2019). Some reports suggested that the expansion of renewable energy has slowed significantly. New wind installations were reported as being the lowest since the beginning of the Energiewende, and it has been suggested that the withdrawal of subsidies for wind undermined the industry, while the new (2017) auction system favors large companies and discourages those community groups which set up their own wind farms and provided a great deal of support for the transition, and possibly made it a democratic process (Buchsbaum, 2019; Morris and Jungjohann, 2016). "By 2012 nearly half of investments in new solar, biomass and wind power came from citizens and energy co-ops" (Morris and Jungjohann, 2016: 8). However, only 35 turbines were installed in the first half of 2019 (Wehrmann, 2019). The AFP (2019) claims that "[m]ore than 600 citizen initiatives have sprung up against the giant installations," which further delays implementation, and that at least one German turbine company is closing, due to shrinking revenue. A wind power summit called by Federal Minister of Economics Peter Altmaier to deal with these problems is still negotiating about actions to take (as of September 2019). Germany was also the world's biggest importer of gas for primary energy, something that the Russian gas pipeline will undoubtedly increase, at the same time increasing energy insecurity and maintaining emissions.

Another problem was revealed in a report by the Federal Agency for Nature Conservation in April 2019, although it was known since at least 2008 (Berg et al., 2008/2010). Apparently to support German companies and keep them competitive, especially during the Energiewende, the German government gives about €55 billion in tax breaks to its biggest polluting industries, through exemptions from levies on kerosene, diesel, and other sources of energy. For example, large corporations such as BASF SE and Thyssenkrupp AG, with very high levels of electricity consumption, are exempted from the Erneuerbare-Energien-Gesetz surcharge (Dulaney, 2019). There is similarly an exemption from tax on diesel usage in agriculture and even subsidies for coal. These subsidies and tax breaks are likely to undermine, or contradict, the function of the Emissions Trading Scheme, or of any other economic factors, in persuading companies not to use fossil fuels or become energy efficient. It could be suggested that, in a neoliberal regime, if these companies have to pay the cost for not using low emissions energy, they can simply go elsewhere, and this is morally justified given neoliberal thinking. It is also difficult for the German government, who could consequently fear severe effects on the German economy if the subsidies were removed. There are, apparently, increasing objections to these subsidies and breaks, but it is by no means certain how the objections will play out politically against such established powers.

Despite these challenges and conflicts, the Coal Commission, issued its final report in January 2019, which changed things yet again (https://www.bmwi.de/Redaktion/EN/Publikationen/commission-on-growth-structural-change-and-employment.html). The report outlines the agreement between representatives of the country's industrial sector, environmental NGOs, citizen initiatives, and policy-makers in a 336-page document that also includes detailed lists of projects for structural change in the affected regions (Cleanenergy

2019; https://www.cleanenergywire.org/factsheets/german-commission-proposes-coal-exit-2038). Out of the 28 official members, 27 voted for the deal. Only the representative of the villages threatened by lignite mine extensions in Lusatia voted against the agreement, saying there were no assurances that the villages would be allowed to remain, while Greenpeace dissented on the exit date, saying it was too late. Niklas Heiland of Oxfam Germany predicted that "If the German government implements the commission's proposal in this way, it is unlikely that the climate goals for 2030 will be achieved" (Schulz, 2019a).

The main recommendation of the commission was to phase out lignite coal mining by 2038. If conditions allow, this could be brought forward to 2035. The commission suggests this option be assessed by 2032. In a first step, it suggests that Germany should switch off 12.5 GW of capacity by 2022. The document gives detailed consideration of how the country can cope with the economic implications that a coal exit will have on the future of mining regions, on the power price, industrial competitiveness, supply security, and the transition to a clean energy system (Cleanenergy, 2019). Overall, news media estimated that affected regions should get some 40 billion euros of support over the next 20 years. The German government has not yet adopted the recommendations of the Coal Commission (as of September 2019). However, all political observers are confident that the Government will approve these recommendations and confirm the exit date for coal mining and production in Germany to be 2038 at the latest. The Bavarian state government has proposed that Germany should stop burning coal by 2030, 8 years earlier than the recommended dates, but it only has 5 out of Germany's 125 coal power plants (Parkin and Wilkes, 2019a). Currently, some people are claiming that cheap gas will render coal and lignite plants unprofitable before this date (Wilkes and Parkin, 2019; Capion, 2019). Leipzig also plans to replace lignite for heating with gas by 2022 (Franke, 2019).

At the same time, the environmental ministry is pushing for a new climate protection act (https://www.bmu.de/themen/klima-energie/klimaschutz/nationale-klimapolitik/klimaschutzgesetz/) that would oblige Germany to be climate neutral over the entire electricity generating system and have very ambitious goals, close to zero emissions, in the area of heating and transport. The details of such a climate protection act are still under debate but there is little doubt that Germany will sharpen its energy regulation so as to reduce CO_2 emission in the energy field to less than 10% of its 1990 volume once it reaches the year 2050. One study has suggested that two population centers in Germany, Mecklenburg-Vorpommern and Schleswig-Holstein, were already close to 100% renewable electricity generation (Brown et al., 2018).

The success of the policies aimed at the envisioned CO_2 reduction will also depend on the implementation of improved energy efficiency measures. Until now, measures aiming at energy efficiency improvements have been predominantly voluntary and based on market incentives and subsidies. The Energy Efficiency Program of the Government (EEF) has announced the goal of reducing primary energy consumption by 20% by 2020 and 50% by 2050 (when compared with 2008). Fig. 2.4 shows the rates of energy efficiency from 2000 to 2015.

Germany is one of world leaders in energy efficiency. In a comparative study of 16 countries, Germany was rated best on the ratio between energy consumption and economic performance (ACEEE, 2015). However, the government has more ambitious goals. There is an increasing pressure on the government to initiate new and binding laws since the national goals for efficiency have not been reached over the last 2 years (AG Energiebilanzen, 2019).

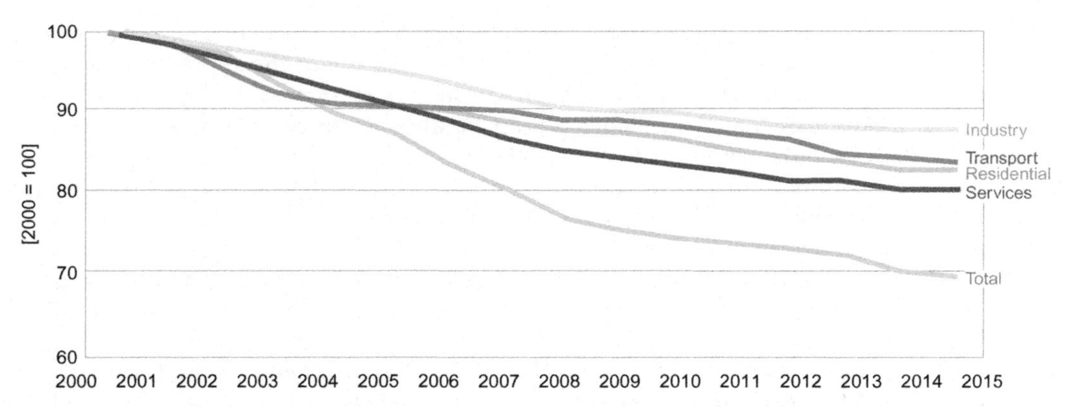

FIGURE 2.4 Technical energy efficiency index. Source: *Adapted from AG Energiebilanzen 2019.*

2.6 Public attitudes toward the energy transition and the request for more participation

The Energiewende aims to promote a transition from a society that covers its energy needs predominantly with fossil fuels, to a "society with long-term secure and sustainable energy supply" (Ethics Commission, 2011). By the middle of this century, Germany intends to emit at least 80% less GHGs than in the reference year 1990 and reduce its primary energy consumption by half when compared with 2008 as a contribution to climate and resource conservation. In the aftermath of the Paris Agreement, Germany wants to accomplish a climate-neutral energy supply system.

Given these ambitious goals, it is not surprising that the energy transition implies a transformation of society as a whole, which has both intra- and inter-generational effects, and that demands a well-coordinated and structured interaction between societal demand, organizational change, new governance tools, technological opportunities (Grunwald et al., 2016) and attention being paid to the inevitable unexpected and unintended consequences of policies, and people's responses to them. Policies have had to be flexible, adjust to complex changing conditions, and deal with their own contradictions, as we have seen. However, the main variable seems to involve the generation of a positive resonance through the population and the many actors who need to be involved in the implementation of the energy transition.

Recognition of the integration of the societal and complex systemic perspective in the implementation of the energy transition is of great importance here, since the reorganization and expansion of the energy system has far-reaching, and possibly unpredictable, consequences for society and its decision-making authorities. Political decisions must increasingly be made under uncertainty, the interests of different actors must be integrated, technological problems and blowbacks must be recognized, and the values, concerns and worries of the citizens taken up and taken to heart.

For the success of the energy transition, it will be decisive how these social concerns are dealt with. The results of public opinion polls provide a mixed picture (Setton et al., 2017; Setton, 2019): the energy transition is viewed positively by a very high proportion of the population and is firmly recognized as a legitimate objective in all social groups. About 90% of the population in Germany is in favor of the energy transition, across all education, income, and age groups, both in rural areas and cities. The majority of supporters of all parties represented in the Bundestag are in favor of the energy transition. The overwhelming majority of the population (80.3%) also sees a personal connection to the energy transition and sees it as a joint task to which everyone in society—including the interviewees themselves—should make a significant contribution.

Almost two-thirds of the population (63%) agree to a phaseout of coal mining, mainly for reasons of climate protection. The majority of the population in the four federal states with lignite mining (Brandenburg, North-Rhine-Westphalia, Saxony, and Saxony-Anhalt) are also in favor of abandoning coal. However, the proportion of supporters is much lower, when compared with the national average. Almost one-third of the population in these areas would like coal production to continue. This sets up fears of far-right political success in those areas, as parties like the AfD argue that the phaseout of lignite shows the elites making decisions that ignore and harm ordinary Germans (Chase, 2018). One AfD speaker said: "The bitter irony of this is that those who are so severely harmed by this are supposed to pay for the damage caused themselves" (Schulz, 2019b).

This all marks the approval of the energy transition. But there is also another side. Although the approval values for the energy turnaround are at a very high level, the verdict of the population on the implementation of the energy transition is much more negative. Nearly half of the respondents rated the development of the German energy transition in 2018 as negative on the whole. This is a significant increase of 14% points when compared with 2017, in which a slightly positive assessment prevailed. Furthermore, less than one in three (31%) rates, the implementation of the Energiewende as being good. Confidence that the energy transition is orderly and based on a convincing plan is also low. More respondents are of the opinion that the energy transition is chaotic (61%). Perhaps people have unrealistic cultural expectations of easy orderly change or they were not prepared for the mess of dealing with complex and competing interests and effects?

The critical assessment is also predominant when it comes to issues of justice, proximity, cost, and political governance. Particularly pronounced is the skeptical view people have on the subject of costs. Three-quarters of the population (75%) consider the energy transition expensive, only 10% perceive it as economical. More than half of the population (51%) considers the energy transition unfair and only one in five (21%) fair.

Disappointment with cost has been further fueled by substantial price hikes in electricity (Bosch and Peyke, 2011). The German government guarantees a fixed price for feeding solar energy into the national grid for a payoff period of 20 years to encourage the installation of solar panels. This led to a dramatic increase in solar panel installations on private property, and a substantial price rise for electricity since 2012. The additional costs for feeding electricity into the national grid affect all electricity consumers with the exception of energy-intensive industries (Andor et al., 2015). These major increases in electricity prices, which were proposed to encourage renewable energy, may have lowered parts of the German public's enthusiasm for renewables and the Energiewende. This constitutes an example of an unintended consequence of an otherwise successful policy.

These critical judgments are an important indicator that implementation and communication of the new energy policies face high levels of scrutiny by the majority of the population. One of the main attempts to gain more public acceptance and support has involved the increase of public participation and stakeholder involvement on the regional and local level (Radke, 2016), even while community participation in wind is decreasing. Although many energy installations in Germany were introduced with accompanying programs of public involvement and informal participation efforts, the overall success of these efforts is still unclear. Most of the papers in this volume address the experiences and outcomes of such participation projects and their likely impacts in the various political arenas.

Even under the constraints of energy price hikes and perceived problems with the implementation of the energy transition, it seems unlikely that Germany will reconsider phasing-out nuclear energy. Nuclear energy has already been surpassed by renewable energy in terms of installed capacity. There is little doubt that the former 23% share of nuclear energy can be replaced by the growing renewable energy sector, provided the collapse in new wind generation does not continue. Second, the consensual agreement to phaseout nuclear energy by the ethics committee, as well as the overwhelming majority vote in the parliament, demonstrated the commitment to the Energiewende from industry, the major players in science and technology and from environmental groups. One of the best indicators for the irreversibility of this decision is the nuclear industry's reaction in Germany. Siemens, the construction and engineering conglomerate that had built all 17 of Germany's nuclear power plants, announced in September 2011 that it would no longer build nuclear power plants anywhere in the world, dropping plans to work with the Russian Rosatom State Atomic Energy Corporation to build new plants (World Energy Council WEC, 2012). All major political social and civil society actors followed the official policy of the German government and prepared for a non-nuclear future. No organized stakeholder group openly opposed the Energiewende, although many disappointed enthusiasts of nuclear energy tried hard to lobby against the Energiewende through Internet initiatives and mailing campaigns. However, there was little public resonance, let alone support, for these campaigns.

For the future, a more interesting policy question is whether the ambitious promise to reduce the amount of fossil fuel used from 80% today to almost zero in 2050 can be kept. Although Germany is highly committed to combating climate change, a commitment rigorously repeated at the 2015 Paris COP 21 meeting, a substantial number of Germans feel much less concerned about the impact of fossil fuels on climate change than they have done about the safety of nuclear energy. Many Germans feel uneasy with the aim of phasing-out both nuclear energy and fossil fuels, and then relying entirely on renewable energy. Given the recent price hikes, the pressure of the unions to keep lignite burning, and the need for energy security, it is not surprising that the public hesitates to welcome an accelerated phaseout of lignite, which could make those situations worse. This could lead to an electoral challenge, especially if people in lignite producing areas feel abandoned. However, with the recommendations of the Coal Commission, it is very likely that the coal era in Germany will come to an end before 2040, provided there are no unexpected political events or disasters.

2.7 Conclusions and policy implications

This article has described the emergence of energy policies in Germany, highlighting the major transitions over the last five decades. Energy policy in Germany reflects the complexity of Germany's political system, the responses of people to contingent events and the uncertain or unintended consequences of actions. The principles of this system are strongly shaped by checks, balances, and conflicts between the States and the Federal government and between the parliament and the courts (Hatch, 1986). There is also a strong corporatist/institutional element in German policy arenas that allows substantial influence from major societal actors such as employers, unions, and, since the 1970s, environmental groups. Policy changes and administrative measures are further subject to judicial review and public debate. This adds complexity, but also adds to the hearing of different views, and public participation.

The sensitivity of politics to public pressure in Germany has made governments vulnerable to grass-root opposition and street protests (Wagner, 1994). As the public in Germany has been more critical of, and active against, nuclear power than in most other nations, the Government has responded quickly to demands for improved safety and control. At the same time, there is a strong pressure from environmental groups to reduce the use of fossil fuel, in particular oil and coal. Pressures from disasters, overseas suppliers, and EU policies also have an effect.

Based on these factors, a major revision of German energy policy has occurred, which included the phasing-out of nuclear power, the expansion of renewable energies, and the introduction of a coal reduction plan (Kern and Löffelsend, 2003). The new policies aim to phaseout nuclear power by 2021, with fossil fuel consumption reduced to at least 20%, if not 0%, of the 1990 level by 2050. After the Paris accord of 2015, the German government has promised to examine the possibility of not using fossil fuel for electricity production at all.

While hardly any influential group in Germany advocates for the continuation of nuclear power, there is still a debate about the future role of lignite in the energy mix. The first argument is that coal energy is still a domestic energy source and the industry employs roughly 120,000 workers (Lauber and Jacobsson, 2015). Local communities around coal mines as well as the powerful mining union can continue to advocate for a base level of coal production and consumption as securing energy security and supporting local economies (Moeller et al., 2014), while other communities can protest against the expansion of mines and the relocation of villages and the destruction of farms. Hardly anyone overtly questions the need for reducing coal consumption but the timing of the phaseout can appear too ambitious, with some advocating a slower and more socially adjusted exit from coal.

The second argument in favor of lignite relates to the volatility of renewable energy. To secure energy security, backup facilities are necessary. Since most coal-fired power plants are already fully financed, they could be economically sustained in an operating mode as such backup. Originally, the plan was to replace coal by gas but it turned out cheaper to keep old coal and lignite-fired power plants than to build new gas-powered plants (Hohmeyer and Bohm, 2015). This, together with the phaseout of nuclear energy, is one of the main reasons why CO_2 emissions increased from 2011 to 2014 even though the share of renewable energy dramatically increased from 17% to 26% of the electricity production. Difficulty of planning is intensified by the uncertain future requirements in a changing system.

The third argument relates to the enormous subsidies put into the energy market. In 2014, German electricity customers paid more than 27 billion Euros to fill the gap between market price and the guaranteed price of renewable energy production (Chrischilles and Bardt, 2016). Many economic analysts believe that Germany will not be able to afford these amounts in the coming years. Coal advocates in particular have embraced this argument to suggest that the increased use of lignite would reduce the financial burden on electricity consumers. These arguments imply that climate change, itself, would have lower costs on the populace, which may not be accurate, but cannot be guaranteed.

Lastly, the new energy situation in Germany has led to a new form of unequal distribution of income and opportunities (González-Eguino, 2015). While more than 40,000 private producers of solar energy can enjoy a guaranteed income from selling electricity that is far greater than present interest rates on capital, poorer sections of society have to pay for these guaranteed prices. Average household electricity prices have risen to three times the level of 2010 prices (Andor et al., 2015). This has produced some disenchantment with the Energiewende and nostalgia for the "golden age" of coal. The income inequalities associated with economic "liberalization" did not produce similar policy concern among political and business groupings, no doubt because they increased "normal" inequalities rather than produced new ones. Whatever one's view on this issue, there is no evidence that further liberalization would, by itself, fix the problem of GHG emissions.

A clear message for policy-makers from this history is the necessity of being prepared to attend to the paradoxical effects and unintended results of policy. This includes the forest dieback from the early attempts to clean coal, the increase in lignite consumption resulting from the success of renewables and decisions based on markets, the increase of CO_2 emissions from 2011 to 2015 (since then they remain more or less stable over all the energy sectors), the unpopular price increases brought about by attempts to popularize renewables, the conflicts over villages and farm lands destroyed for lignite mines, and complaints over lack of citizen participation. Contingent events such as technological accidents (Chernobyl and Fukushima) and time limits appear to affect decision-making processes, and some corporate interests seem resilient and able to take advantage of change (even if to slow that change) while others do not. Furthermore, liberalization easily changes into the subsidy of powerful economic actors, which may work against the transition's effectiveness, as with the early stages of the ETS, the maintenance of diesel as a fuel, or the inhibition of further community energy projects. The European context can also make policy carryout more complicated, due to continent-wide regulation, and conflict between countries with different needs and histories.

Interestingly enough, carbon sequestration (CCS) has never gained much popularity in Germany. After local communities protested against the first experimental storage facilities in northern Germany, the idea of CCS has lost attractiveness among political parties as well as stakeholders (Dütschke et al., 2015). It must be seen, if in the aftermath of the 2015 Paris accord, CCS might get a second chance in Germany, despite its acknowledged problems (cf Marshall, 2016).

At this current moment, it is unclear how the energy supply system will finish up, as the transition continues to produce surprising results. It is, however, unlikely that Germany will revoke its antinuclear policy but it is still not clear if, in light of growing costs and the inability to compensate for volatile energy generation from renewable sources, whether lignite or gas will experience a longer presence in Germany's energy generation and consumption than presently seems envisioned by the German government. It is clear that the share of coal will decrease and the 2019 recommendations of the Coal Commission argued for a phaseout before 2038. Over time, this goal could undergo serious revision in both directions (earlier or later). It is clear that coal will play a significant role in the interim, but no politically powerful group (outside of the coal companies) is advocating a reinvestment in coal. The timing may be slower than requested by most environmental groups, but a golden era of coal is unlikely to reappear.

Overall, Germany seems to be on track for completing the energy transition as promised in the next 30 years. Much of the success will depend on the capability and willingness of political and economic authorities to provide better and more effective forms of public engagement and participation in the implementation of the energy transition (Renn, 2019). As much as the German population approves the transformation toward renewable energy sources, they are very skeptical about how the transition has been implemented, about its costs, and how regional and local actors are being involved. This is the Achilles heel of the energy transition. If the Energiewende retains its momentum and gains public support even at different times, it could act as a role model for many other countries in the world that sympathize with a phaseout of both nuclear and coal energy supply, but which are reluctant because of economic fears. Should Germany not succeed, then the era of nuclear and fossil energy may continue, despite the increasing risk of climate turmoil.

Acknowledgments

Work by Jonathan Marshall was funded by an Australian Research Council Future Fellowship FT160100301 "Society and climate change: A social analysis of disruptive technology." Work by Ortwin Renn is funded by the Institute for Advanced Sustainability Studies (IASS) in Potsdam, Germany, and was conducted as part of the Kopernikus Project "System Integration: Energiewende Navigation System (ENavi)" funded by the Federal Ministry for Education and Research, Germany.

References

ACEEE, 2015. The International Energy Efficiency Scorecard. American Council for an Energy-Efficient Economy, Washington, DC. Available from: http://aceee.org/portal/national-policy/international-scorecard.

AFP, 2019. Turbulent politics: how wind energy became a divisive issue in Germany. thelocal. de 5 September 2019. <https://www.thelocal.de/20190905/turbulent-politics-how-wind-energy-became-a-divisive-issue-in-germany>.

AG Energiebilanzen, 2019. Energie. Daten und Fakten. <https://ag-energiebilanzen.de/2-0-Daten-und-Fakten.html>.

Andor, M., Frondel, M., Schmidt, C.M., Simora, M., Sommer, S., 2015. Klima-und Energiepolitik in Deutschland–Dissens und Konsens. List. Forum für Wirtschafts-und Finanzpolit. 41 (1), 3–21.

Auge, L., 2019. Ursula von der Leyen Must Take on Europe's biggest #Climate rebel: Germany. EUreporter July 25, 2019. <https://www.eureporter.co/frontpage/2019/07/25/ursula-von-der-leyen-must-take-on-europes-biggest-climate-rebel-germany/>.

Berg, H., Burger, A., Thiele, D., 2008/2010. Environmentally Harmful Subsidies in Germany. Federal Environment Agency. Available from: https://www.umweltbundesamt.de/sites/default/files/medien/publikation/long/3896.pdf.

Beuermann, C., Jaeger, J., 1996. Climate change politics in Germany: how long will any double dividend last? In: O'Riordan, T., Jaeger, J. (Eds.), Politics of Climate Change: A European Perspective. Routledge, London and New York, pp. 186–226.

BMU, 2019. Klimaschutzbericht 2018. <https://www.bmu.de/download/klimaschutzbericht-2018/>.

BMWE, 2016. Zahlen und Fakten. Bundesministerium für Wirtschaft und Energie. <http://www.bmwi.de/DE/Themen/Energie/Strommarkt-der-Zukunft/zahlen-fakten.html>.

Boehmer-Christiansen, S.A., 1992. Taken to the cleaners: the fate of the East German energy sector since 1990. Environ. Politics 1 (2), 196–228.

Boren, Z., 2018. How Germany Quietly Turned Against Action on Climate Change. Unearthed 03.10.2018. <https://unearthed.greenpeace.org/2018/10/03/how-germany-quietly-turned-against-action-on-climate-change/>.

Bosch, S., Peyke, G., 2011. Gegenwind für die Erneuerbaren – Räumliche Neuorientierung der Wind-, Solar- und Bioenergie vor dem Hintergrund einer verringerten Akzeptanz sowie zunehmender Flächennutzungskonflikte im ländlichen Raum. Raumforsch. und Raumordn. 69 (2), 105–118.

Brauers, H., Herpich, P., von Hirschhausen, C., Jürgens, I., Neuhoff, K., Oei, P.-Y., et al., 2018. Coal transition in Germany - learning from past transitions to build phase-out pathways. IDDRI and Climate Strategies. Paris and Berlin.

Brown, T.W., Bischof-Niemz, T., Blok, K., Breyer, C., Lund, H., Mathiesen, B.V., 2018. Response to 'Burden of proof: a comprehensive review of the feasibility of 100% renewable-electricity systems'. Renew. Sustain. Energy Rev. 92, 834–847.

Bruhns, H., Keilhacker, M., 2011. Energiewende Wohin führt der Weg? Aus Polit. und Zeitgesch. 46-47, 22–29.

Bryant, G., 2016. Creating a level playing field? The concentration and centralisation of emissions in the European Union Emissions Trading System. Energy Policy 99, 308–318.

Buchan, D., Malcolm, K., 2016. Europe's Long Energy Journey: Towards an Energy Union? OUP Catalogue. Oxford University Press, NO. 9780198753308.

Buchholz, W., 2011. Energiepolitische Implikationen einer Energiewende. Ifo-TUM Symposium zur Energiewende in Deutschland. Manuskript.

Buchsbaum, L.M., 2019. German renewable energy cooperatives struggle as markets collapse. Energytransition.org, June 19, 2019, <https://energytransition.org/2019/06/german-renewable-energy-cooperatives-struggle-as-markets-collapse/#more-19973>.

Capion, K., 2019. Why German coal power is falling fast in 2019. RenewEconomy, July 22, 2019, <https://reneweconomy.com.au/why-german-coal-power-is-falling-fast-in-2019>.

Chase, J., 2018. Eastern German states demand €60 Billion for coal phaseout. DW October 19, 2018. <https://www.dw.com/en/eastern-german-states-demand-60-billion-for-coal-phaseout/a-45961437>.

Chrischilles, E., Bardt, H., 2016. Fünf Jahre nach Fukushima: Eine Zwischenbilanz der Energiewende. IW-Report, No. 6/2016, Institut der deutschen Wirtschaft (IW), Köln.

DG Energy, 2011. After Fukushima: EU Stress Tests Start on 1 June. EU Commissioner for Energy, Press Release IP/11/640, Brussels, 25 May.

Dulaney, C., 2019. Germany Looks to Revamp Billions in Tax Subsidies for Polluters. Bloomberg, Sept. 5. <https://news.bloombergtax.com/daily-tax-report-international/germany-under-pressure-to-overhaul-billions-in-tax-breaks-subsidies-for-polluters-1>.

Dütschke, E., Schumann, D., Pietzner, K., 2015. Chances for and limitations of acceptance for CCS in Germany. In: Liebscher, A., Münch, U. (Eds.), U. (eds.): Geological Storage of Co2: Long Term Security Aspects. Springer, Heidelberg, pp. 229–245.

Energywire, 2019. How Germany's Coal Deal Affects E.U.'s Paris Target. <https://www.eenews.net/climatewire/stories/1060119251>.

Ethics Commission, 2011. Deutschlands Energiewende. Ein Gemeinschaftswerk für die Zukunft. Endbericht. Berlin.

EU, n.d. Treaty establishing the European Coal and Steel Community, ECSC Treaty. <https://eur-lex.europa.eu/legal-content/EN/TXT/HTML/?uri = LEGISSUM:xy0022&from = EN>.

Franke, A., 2019. Germany's Leipzig city to exit lignite-fired heating by 2022. SPG, July 9, 2019. <https://www.spglobal.com/platts/en/market-insights/latest-news/coal/070919-germanys-leipzig-city-to-exit-lignite-fired-heating-by-2022>.

Friedmann, W., 1955. The European coal and steel community. Int. J. 10 (1), 12–25.
González-Eguino, M., 2015. Energy poverty: an overview. Renew. Sustain. Energy Rev. 47, 377–385.
Goutte, S., Vassilopoulos, P., 2019. The value of flexibility in power markets. Energy Policy 125, 347–357.
Grunwald, A., Renn, O., Schippl, J., 2016. Fünf Jahre integrative Forschung zur Energiewende: Erfahrungen und Einsichten. GAIA - Ecol. Perspect. Sci. oc. 25 (4), 302–304.
Guisan, C., 2011. From the European coal and steel community to Kosovo: reconciliation and its discontents. J. Common. Mark. Stud. 49 (3), 541–562.
Hatch, M.T., 1986. Politics and nuclear power - energy policy in Western Europe. The University Press of Kentucky, Lexington.
Hohmeyer, O.H., Bohm, S., 2015. Trends toward 100% renewable electricity supply in Germany and Europe: a paradigm shift in energy policies. Wiley Interdiscip. Reviews: Energy and. Environ. 4 (1), 74–97.
International Energy Agency, 2012. Oil and gas security. Germany. IEA: Paris. <http://www.iea.org/publications/freepublications/publication/germanyoss.pdf>.
International Energy Agency (IEA), 2007. Energy Policies of IEA Germany. 2007. IEA: Paris.
Kenk, G., Fischer, H., 1988. Evidence from nitrogen fertilisation in the forests of Germany. Environ. Pollut. 54 (3), 199–218.
Kern, K.K., Löffelsend, T., 2003. Die Umweltpolitik der rot-grünen Koalition - Strategien zwischen nationaler Pfadabhängigkeit und globaler Politikkonvergenz. WZB Discussion Paper. SP IV 200-103. Berlin.
Kleinert, T., 2011. Change 2022: Der Atomausstieg: Chancen und Risiken der erneuerbaren Energien unter Berücksichtigung vornehmlich ökonomischer Daten. Grin Verlag: Norderstedt.
Kuebler, M., 2019. Blocked EU fails on climate pledge, despite increasing public pressure. DW 21.06.2019. <https://www.dw.com/en/blocked-eu-fails-on-climate-pledge-despite-increasing-public-pressure/a-49291654>.
Lauber, V., Jacobsson, S., 2015. Lessons from Germany's Energiewende. In: Fagerberg, J., Laesradius, S., Martin, B.R. (Eds.), The Triple Challenge for Europe: Economic Development, Climate Change, and Governance. University Press, Oxford, UK, pp. 173–203.
Marshall, J.P., 2016. Disordering fantasies of coal and technology: carbon capture and storage in Australia. Energy Policy 99, 288–298.
Martinez, C., Byrne, J., 1996. Science, society and the state: The nuclear project and the transformation of the American political economy. In: Byrne, J., Hoffman, S.M. (Eds.), T: Governing the Atoms - the Politics of Risk. Transaction Publishers, New Brunswick, London.
Matlary, J.H., 1997. Energy Policy in the European Union. MacMillan Press Ltd, Basingstoke, London.
McCollum, D., Bauer, N., Calvin, K., Kitous, A., Riahi, K., 2014.): Fossil resource and energy security dynamics in conventional and carbon-constrained worlds. Climatic Change 123 (3-4), 413–426.
Meza, E., 2013. Solar Self-Consumption on the Rise in Germany. PV Magazine. <http://www.pv-magazine.com/news/details/beitrag/solar-self-consumption-on-the-rise-in-germany_100012123/#axzz3vX1GFZK1>.
Moeller, C., Meiss, J., Mueller, B., Hlusiak, M., Breyer, C., Kastner, M., et al., 2014. Transforming the electricity generation of the Berlin–Brandenburg region, Germany. Renew. Energy 72, 39–50.
Monbiot, G., 2017. The Smog Chancellor. 21st September 2017 [Originally in The Guardian 20th September 2017]. <https://www.monbiot.com/2017/09/21/the-smog-chancellor/>.
Morris, Jungjohann, 2016. Energy Democracy: Germany's Energiewende to Renewables. Palgrave Macmillan.
Morton, T., Müller, K., 2016. Lusatia and the coal conundrum: The lived experience of the German Energiewende. Energy Policy 99, 277–287.
Oberthür, S., 2019. Hard or soft governance? The EU's climate and energy policy framework for 2030. Politics Gov. 7 (1). Available from: https://doi.org/10.17645/pag.v7i1.1796.
Parkin, B., Wilkes, W., 2019a. Bavaria seeks fast-track german coal exit in snub to merkel plan. Bloomberg June 24, 2019. <https://www.bloomberg.com/news/articles/2019-06-24/bavaria-seeks-fast-track-german-coal-exit-in-snub-to-merkel-plan>.
Parkin, B., Wilkes, W., 2019b. Merkel plans broad emissions levy in last major climate push. Bloomberg, July 12, 2019. <https://www.bloomberg.com/news/articles/2019-07-12/merkel-plans-broad-emissions-levy-in-last-major-climate-push>.
Radke, J., 2016. Bürgerenergie in Deutschland. Partizipation zwischen Gemeinwohl und Rendite. Springer, Wiesbaden.
Reizenstein, A., Kapop, R., 2019. The German coal commission. A role model for transformative change? ESG Briefing Paper. <https://www.e3g.org/docs/E3G_2019_Briefing_German_Coal_Commission.pdf>.
Renn, O., 2008. Risk Governance. Coping with Uncertainty in a Complex World. Earthscan, London.

Renn, O., 2019. Inter- und Transdisziplinäre Forschung: Konzept und Anwendung auf die Energiewende. Angew. Philosophie (Appl. Philosophy). Heft 1 (2019), 54–75.

Renn, O., Dreyer, M., 2013. Risk Governance: Ein neues Steuerungsmodell zur Bewältigung der Energiewende. In: Vogt, M., Ostheimer, J. (Eds.), Die Moral der Energiewende. Risikowahrnehmung im Wandel am Beispiel der Atomenergie. Kohlhammer, Stuttgart.

Roy, N., 2019. Germany denies being reliant on Russian energy supplies – states importance of Nord Stream 2. Peace Times July 9, 2019. <https://dailypeacetimes.com/2019/07/09/germany-denies-being-reliant-on-russian-energy-supplies-states-importance-of-nord-stream-2>.

Rueter, G., 2017. How Green is Angela Merkel? DW. 18.09.2017. <https://www.dw.com/en/how-green-is-angela-merkel/a-40565741>.

Sanderson, F., 1958. The Five-Year Experience of the European Coal and Steel Community. Int. Organ. 12 (2), 193–200.

Schauenberg, T., 2019. As Germany phases out coal, villages still forced to make way for mining. DW. March 22, 2019. <https://www.dw.com/en/as-germany-phases-out-coal-villages-still-forced-to-make-way-for-mining/a-48017253>.

Schirrmeister, M., 2014. Controversial Futures—Discourse Analysis on Utilizing the "Fracking" Technology in Germany. Eur. J. of. Futures Res. 2 (1), 1–9.

"Schulz, F., 2019a. What's in the German coal commission's final report? EURACTIV.de Jan 29 2019. <https://www.euractiv.com/section/energy/news/whats-in-the-german-coal-commissions-final-report/>.

Schulz, F., 2019b. German Coal Phase-Out Criticized but Welcomed on Whole. EURACTIV.de Jan 30. 2019. <https://www.euractiv.com/section/energy/news/wider-german-public-welcomes-compromise-for-phasing-out-coal/>.

Schumann, R., 1950. The Schuman Declaration – 9 May 1950. <https://europa.eu/european-union/about-eu/symbols/europe-day/schuman-declaration_en>.

Setton, D., 2019. Soziales Nachhaltigkeitsbarometer der Energiewende 2018. Kernaussagen und Zusammenfassung der wesentlichen Ergebnisse. - IASS Study, February 2019. Available from: https://doi.org/10.2312/iass.2019.002.

Setton, D., Matuschke, I., Renn, O., 2017. Soziales Nachhaltigkeitsbarometer der Energiewende 2017: Kernaussagen und Zusammenfassung der wesentlichen Ergebnisse. Institute for Advanced Sustainability Studies: Potsdam. Available from: https://doi.org/10.2312/iass.2017.019.

Stam, C., 2019. 'Merkel deceived us on climate', says Germany's Energiewende godfather EURACTIV, Jan 15, 2019. <https://www.euractiv.com/section/energy/news/merkel-deceived-us-on-climate-says-germanys-energiewende-godfather>.

Volker-Quaschning, 2015. Statistiken: Deutschland versagt beim Klimaschutz: Treibhausgasemissionen steigen 2015 wieder an. <http://volker-quaschning.de/datserv/CO2-spez/index.php>.

Wagner, P., 1994. Contesting Policies and Redefining the State: Energy Policy-making and the Anti-nuclear Movement in West Germany. In: Flam, H. (Ed.), States and Anti-nuclear Movements. Edinburgh University Press, Edinburgh, pp. 264–298.

Wehrmann, B., 2019. Germany's onshore wind power expansion threatens to grind to a halt. Cleanenergywire 25 July 2019. <https://www.cleanenergywire.org/news/germanys-onshore-wind-power-expansion-threatens-grind-halt>.

Wettengel, J., 2019. Germany's dependence on imported fossil fuels. CleanEnergyWire. April 29, 2019. <https://www.cleanenergywire.org/factsheets/germanys-dependence-imported-fossil-fuels>.

Wilkes, W., Parkin, B., 2019. Markets Drive Germany's Exit From Coal Much Harder Than Merkel. Bloomberg. July 5, 2019. <https://www.bloomberg.com/news/articles/2019-07-05/markets-drive-germany-s-exit-from-coal-much-harder-than-merkel>.

World Energy Council (WEC), 2012. World Energy Perspective: Nuclear Energy One Year After Fukushima. WEC, London.

CHAPTER

3

Inclusive governance for energy policy making: conceptual foundations, applications, and lessons learned

Ortwin Renn and Pia-Johanna Schweizer

Institute for Advanced Sustainability Studies (IASS), Berliner Strasse, Potsdam, Germany

OUTLINE

The Role of Public Participation in Energy Transitions
DOI: https://doi.org/10.1016/B978-0-12-819515-4.00003-9

3.1 Introduction

Policy making in the energy sector is subject to numerous challenges and influences. Secure, environmentally friendly, and affordable energy supply for now and future generations constitute the shared basis for policy making. It is, however, not clear what exactly secure, environmentally friendly, and affordable energy supply means, and how it can be accomplished in a policy arena in which different stakeholders hold different views and interests. Safety levels, environmental and climate quality goals, energy prices, and energy production are subject to scientific dissent, normative value claims, and divergent risk perceptions. The debate on energy is fueled by long-lasting conflicts on nuclear power in several countries, with a peak in public opposition in the aftermath of the Fukushima disaster. Many countries, including Germany, decided to phase-out nuclear energy and to revoke licenses for nuclear power plants (Renn and Marshall, in this volume). More recently, the Friday for Future movement has placed climate protection in the center of political activities for replacing fossil fuel by renewable energy sources. Due to this debate, fossil fuels have come under severs attack by environmental groups. The low acceptance and strong opposition to nuclear power as well as the envisioned phase-out of all fossil fuels had led to fundamental changes in official governmental policies, for example, the "Energiewende" in Germany (Lauber and Jacobsson, 2015).

Such a major transition cannot succeed without the support of the major stakeholders in society and the majority of the affected publics. In democratic societies, such fundamental transformations require a public discourse devoted to determine the conditions under which a safe, reliable, environmentally friendly, and affordable energy supply can be designed. Given the multitude of objectives and goals, it is unlikely that solution will emerge that meet all the objectives simultaneously. Rather conflicting values will prevail and painful trade-offs, for example, between climate protection and the preservation of employment in coal mining and coal combustion, need to be negotiated (Brauers et al., 2018).

Such negotiations will take place in national parliaments and executive branches of government. Yet given the need for fundamental changes in infrastructure, regulations and lifestyles, stakeholder, and affected groups of the public demand to be involved in the decision-making and, in particular, implementation process. The objective of involvement in the policy making process is to foster a shared understanding among stakeholders and the general public of how to combine and use different energy sources, how to substitute fossil fuel by renewable energy, and how to reach a higher energy efficiency or even sufficiency in terms of lifestyle and consumption (Heuberger and Mac Dowell, 2018). The overall goal is to design a policy program for a sustainable energy future.

Since such a complex task requires the integration of expertise and public participation (Renn, 2008: 284ff). On the one hand, policy makers need sufficient knowledge about the potential impacts and likely consequences of policy options under investigation. On the other hand, they need criteria to judge the desirability or undesirability of these consequences for the people affected and the public at large (Horlick-Jones et al., 2007; Renn and Schweizer, 2009). Criteria on desirability are reflections of social values such as good health, equity, environmental quality, or efficient use of scarce resources. Both components—knowledge and values—are essential for any decision-making process independent of the issue and the problem context. However, in the energy debate, such an integration is particularly valuable. Neither the concept of technocracy (experts know the best answer and the democratic institutions should at best confirm them) nor that of postfactual arbitrariness (whatever we want we can accomplish it independent of the facts) will resolve the problem (Nanz and Leggewie, 2019: 13ff). In order for societies to move toward more sustainable energy systems, procedural structures are urgently needed that build upon the best available expertise and the informed consent of those who will experience the consequences of the requested changes.

This chapter is an attempt to develop some conceptual and operational guidelines for making stakeholder involvement successful with respect to resolving complex policy problems and generating solutions that are both scientifically sound and ethically acceptable. The guidelines are based on the normative belief that the integration of knowledge and values can best be accomplished by involving those actors in the policy making process that are able to contribute the respective knowledge as well as the variability of values necessary to make effective, efficient, fair, and morally acceptable decisions (Tuler and Webler, 1995; Webler, 1999; IRGC (International Risk Governance Council), 2005; Rosa et al., 2014: 78ff).

Section 3.2 will introduce the two fundamental questions of any participatory decision-making: inclusion and closure. What and who will be in included in the deliberation process and what arrangement can be made for reaching a joint agreement? Section 3.3 will explain the potential contributions that different stakeholders can bring into the negotiation process about risk management. This section will distinguish between economic, political, scientific, and civil society groups that have different types of knowledge and values to integrate into the decision-making process. Section 3.4 explores the various perspectives that are associated with the goals and rationales of stakeholder involvement. The section introdintroduces sixuces perspectives: functional, neoliberal, deliberative, anthropological, and postmodern. Based on the discussion of these five perspectives, Section 3.5 makes an argument for a combination of the functionalist and deliberative perspective: the so-called analytic–deliberative process. Section 3.6 provides some operational guidelines of how to structure analytic–deliberative processes and how to find the best instrument for the given purpose. The section will give an example of an analytic–deliberative discourse and demonstrates how it has worked in practical applications before the last Section 3.7 provides some concluding remarks about risk management and participation.

3.2 Essential questions for organizing participation: inclusion and closure

Anticipating consequences of human actions or events (i.e., knowledge) and evaluating the desirability and moral quality of these consequences (i.e., values) pose particular problems if the consequences are complex and uncertain and the values contested and controversial. Dealing with complex, uncertain and socially contested outcomes often lead to the emergence of social conflict (Renn et al., 2011; Klinke and Renn, 2019). Although everyone may agree on the overall goal of sustainable energy supply, precisely what that goal entails (how sustainable is sustainable enough?) and precisely how that goal will be obtained (who bears the risks and who reaps the benefits?) may evoke substantial disagreement (Hagendijk and Irwin, 2006). Typical questions in this context are as follows: what are the most suitable criteria for judging the appropriateness of the proposed energy policies? What role should the assessment of uncertainty and ignorance play in designing new policies? How should one balance a variety of options with a mix of benefits and costs, thus requiring painful trade-offs to be made. How should society regulate energy systems? How can a social fair distribution of benefits and risks be assured?

These crucial questions of how to deal with complex, uncertain, and controversial topics such as energy policies go beyond the conventional routines. Numerous strategies to cope with this challenge have evolved over time. They include technocratic decision-making through the explicit involvement of expert committees, muddling through in a pluralist society, negotiated rulemaking via stakeholder involvement, deliberative democracy, or ignoring probabilistic information altogether (see review in Renn, 2008: 90ff) The main thesis of this paper is that public and private institutions that assess and design energy policies are in urgent need of revising their institutional routines and of designing procedures, which enable them to integrate professional assessments (systematic knowledge), adequate institutional process (political legitimacy), responsible handling of public resources (efficiency), and public knowledge and perceptions (reflection on public values and preferences). These various inputs require the involvement of different stakeholders as well as representatives of the nonorganized public that will be affected by the policies in their livelihood.

In a pluralist society, knowledge claims about potential consequences of actions as well as criteria for judging the moral acceptability of options are contested. One could assume a benevolent dictator who tries to act in the best interest of the common good or one could delegate the power to decide to elected representatives who can gather the best knowledge from experts and make the decision in correspondence to the interests and values of their respective constituencies. The literature refers to these two modes of collective decision-making as technocratic and decisionistic (Habermas, 1968; Stirling, 2004). A third mode that Habermas has coined "pragmatic" is based on the assumption that the plural actors of society should be an integral part of the policy making process. This participatory mode of decision-making is nowadays referred to as inclusive governance (Jasanoff, 1993; Renn, 2008: 273ff; Renn, 2014; Schweizer et al., 2016).

Inclusion describes the question of what and whom to include into the governance process, not only into the decision-making, but into the whole process from framing the problem, generating options, and evaluating each of these options and in the end arriving at a joint conclusion. This goal presupposes that, at least, major attempts have been made to meet the following conditions (cf. IRGC (International Risk Governance Council), 2005: 49f; Webler, 1999; Wynne, 2002; Renn and Schweizer, 2009):

- Representatives of all relevant stakeholders have been involved (if appropriate).
- All representatives of the various stakeholder groups or directly affected populations have been empowered to participate actively and constructively in the discourse.
- The framing of the risk problem (or the issue) has been codesigned in a dialogue with the different groups.

- A common understanding of the magnitude of the risk and the potential risk-management options has been generated and a plurality of options that represent the different interests and values of all involved parties have been included.
- Major efforts have been made to conduct a forum for decision-making that provides equal and fair opportunities for all parties to voice their opinion and to express their preferences.
- There exists a clear connection between the participatory bodies of decision-making and the corporate or political implementation level.

These objectives can be accomplished in most cases in which energy system are governed on a local level, since the different parties there are familiar with each other and with the energy policy in question. It is, however, much more difficult to reach these objectives for energy decisions that concern actors on a national or global level, and where the policy options are characterized by high complexity, uncertainty, and ambiguity. Sometimes, parties may have an advantage when stalling the entire consultation process, because their interests profit from leaving the existing strategies into place. Consequently, inclusive governance processes need to be thoroughly monitored and evaluated, to prevent such strategic deconstructions of the process.

Closure, on the other hand, is needed to restrict the selection of policy options, to guarantee an efficient use of resources, be it financial or the use of time and effort of the participants in the governance process. Closure concerns the part of generating and selecting decision options, more specifically: Which options are selected for further consideration, and which options are rejected. Closure therefore concerns the product of the deliberation process. It describes the rules of when and how to close a debate, and what level of agreement is to be reached. The quality of the closure process has to meet the following requirements (cf. IRGC (International Risk Governance Council), 2005: 50; Webler, 1995; Renn and Schweizer, 2009):

- Have all arguments been properly treated? Have all truth claims been fairly and accurately tested against commonly agreed standards of validation?
- Has all the relevant evidence, in accordance with the actual state-of-the-art knowledge, been collected and processed?
- Was systematic, experimental, and practical knowledge and expertise adequately included and processed?
- Were all interests and values considered, and was there a major effort to come up with fair and balanced solutions?
- Were all normative judgments made explicit and thoroughly explained? Were normative statements derived from accepted ethical principles or legally prescribed norms?
- Were all efforts undertaken to preserve plurality of lifestyle and individual freedom and to restrict the realm of binding decisions to those areas in which binding rules and norms are essential and necessary to produce the wanted outcome?

If these requirements are met, there is at least a chance of being able to achieve an agreement at the end of the deliberation process. The success of the stakeholder and public involvement strongly depends on the quality of the process. Consequently, this process has to be specifically designed for the context and characteristics of policies under review. The balance of inclusion and closure is one of the crucial tasks of effective energy governance.

3.3 The need for stakeholder involvement in risk governance

3.3.1 A systems analytic view on society

At the foundation of any society is the need for *effectiveness*, *efficiency*, *resilience*, and *social cohesion* (Renn, 2014: 533; Rosa et al., 2014: 173ff). *Effectiveness* refers to the need of societies to have a certain degree of confidence that human activities and actions will actually result in the consequences that the actors intended when performing them. *Efficiency* describes the degree to which scarce resources are used to reach the intended goal. The more resources that are invested to reach a given objective, the less efficient is the activity under question. *Resilience* describes the capacity to sustain functionality of a system or a service even under severe stress or unfamiliar conditions. Finally, *social cohesion* covers the need for social integration and collective identity despite plural values and lifestyles (Parsons, 1951, 1971). All four needs or functions of society build the foundation for legitimacy. *Legitimacy* is a composite term that denotes, first, the normative right of a decision-making body to impose a decision even on those who were not part of the decision-making process (issuing collectively binding decisions), and second, the factual acceptance of this right by those who might be affected by the decision (Suchman, 1995). As a result, it includes an objective normative element, such as legality or due process, and a subjective judgment, such as the perception of acceptability (Luhmann, 1983).

Within the macro-organization of modern societies, these four functions are predominantly handled by different societal systems: economy, science (expertise), politics (including legal systems), and the social sphere (Münch, 1982; Parsons, 1967; Rosa et al., 2014: 175). In the recent literature on governance the political system is often associated with the rationale of hierarchical and bureaucratic reasoning; the economic system with monetary incentives and individual rewards; and the social sphere with the deregulated interactions of groups within the framework of a civil society.[1] Another way to phrase these differences is by distinguishing among competition (market system), hierarchy (political system), and cooperation (sociocultural system).

Scientific input into these models is seen as an integral part of politics in the form of scientific advisory committees or the civil sector in the form of independent institutions of knowledge generation (Alexander, 1993). In the field of decision-making about risks, we prefer to operate with four separate subsystems since scientific expertise is crucially important to risk decisions and cannot be subsumed under the other three systems (see also cf. Joss, 2005: 198ff). Since science is part of a society's cultural system, our proposed

[1] In this text, we use the definition of civil society put forward by Alexander (1993: 797): "the realm of interaction, institutions and solidarity that sustains public life of societies outside the worlds of economy and state."

classification is also compatible with the classic division of society into four subsystems (politics, economics, culture, and social structure), which has been suggested by the functional school in sociology (Parsons and Shils, 1951; Münch, 1982). The picture of society is, of course, more complex than the division into four systems suggests. Many sociologists relate to the concept of "embeddedness" when describing the relationships among the four systems (Granovetter, 1992). Each system is embedded in the other systems and mirrors the structure and functionality of the other systems in subsystems of their own. To make our argument, however, the simple version of four analytically distinct systems is sufficient.

Each of the four systems is characterized by several governance processes and structures adapted to the system properties and functions in question. The four systems and their most important structural characteristics are shown in Fig. 3.1. What findings can be inferred from a comparison of these four systems?

In the market system, decisions are based on the cost–benefit balance established on the basis of individual preferences, property rights, and individual willingness to pay. The conflict resolution mechanisms relate to civil law regulating contractual commitments, Pareto optimality (each transaction should make at least one party better off without harming third parties), and the application of the Kaldor–Hicks criterion (if a third party is harmed by a transaction, this party should receive financial or in-kind compensation to such an extent that the utility gained through the compensation is at least equivalent to the disutility experienced or suffered by the transaction). The third party should hence be at least indifferent between the situation before and after the transaction. In economic theory the transaction is justified if the sum of the compensation is lower than the surplus that the parties could gain as a result of the planned transaction. However, the compensation does not need to be paid to the third party. Additional instruments for dealing with conflicts are (shadow) price setting, the transfer of rights of ownership for public or nonrival goods, and financial compensation (damages and insurance) to individuals whose utilities have been reduced by the activities of others. The main goal here is to be efficient.

In politics, decisions are made on the basis of institutionalized procedures of decision-making and norm control (within the framework of a given political culture and system of government). The conflict resolution mechanism in this sector rests on due process and procedural rules that ideally reflect a consensus of the entire population. In particular, decisions should reflect the common good and the sustainability of vital functions to society. This is why resilience lies at the heart of public activities. In democratic societies the division in legislative, executive, and judicial branches; defined voting procedures; and a structured process of checks and balances lie at the heart of the institutional arrangements for collective decision-making. Votes in a parliament are as much a part of this governance model as is the challenging of decisions before a court. The target goal here is to seek resilience as a major prerequisite of legitimacy.

Science has at its disposal methodological rules for generating, challenging, and testing knowledge claims, with the help of which one can assess decision options according to their likely consequences and side effects. If knowledge claims are contested and conflicts arise about the validity of the various claims, scientific communities make use of a wide variety of knowledge-based decision methods, such as methodological review or retests,

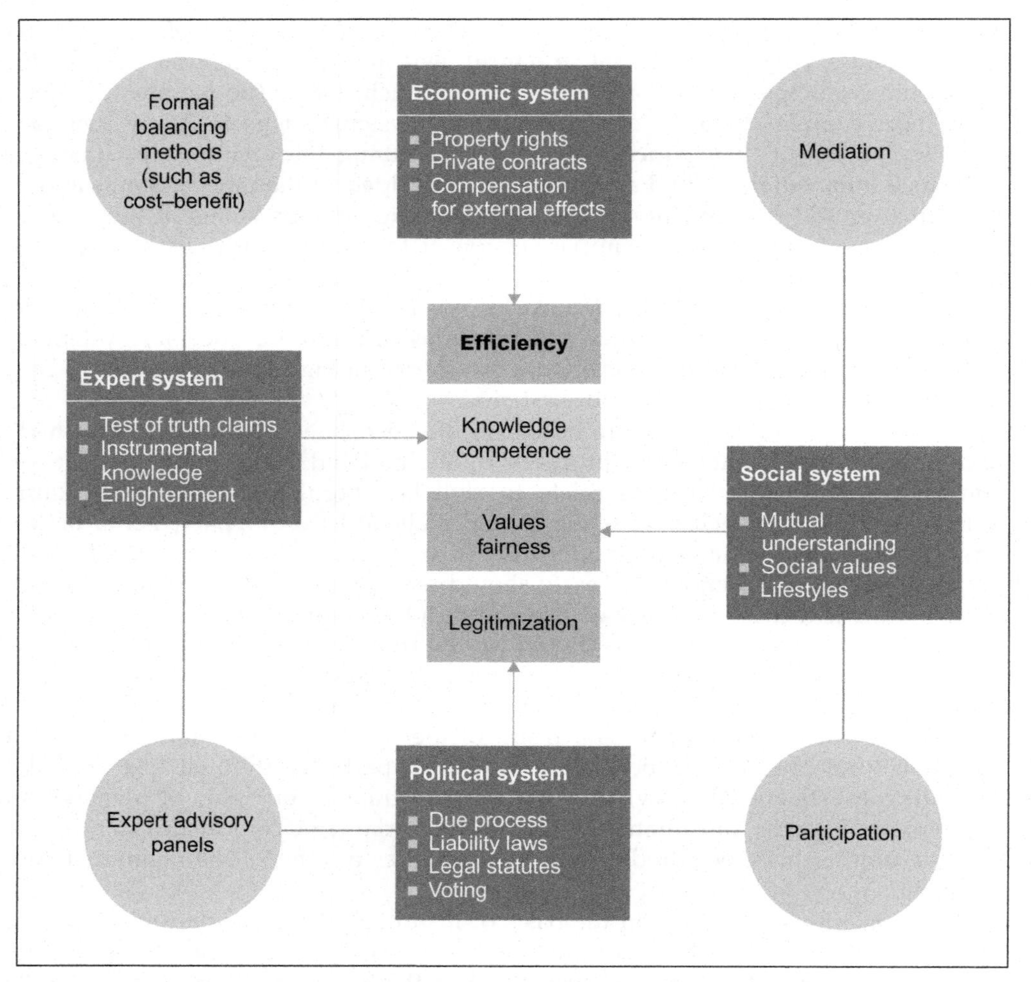

FIGURE 3.1 Four central systems of society. Source: *Based on Münch 1982; adapted from Renn, O., 2008: Risk Governance. Coping With Uncertainty in a Complex World. Earthscan, London, p. 287.*

metaanalysis, consensus conferences, Delphi, or (most relevant in this arena) peer review to resolve the conflicts and test the explanatory or predictive power of the truth claims. These insights help policy makers understand phenomena and be effective in designing policies.

Finally, in the social system, there is a communicative exchange of interests, preferences, and arguments assisting all actors to arrive at a unanimous solution. Conflicts within the social system are normally resolved by finding favorable arrangements for all parties involved, using empathy as a guide to explore mutually acceptable solutions, referring to mutually shared beliefs, convictions, or values or relying on social status to justify one's authority. These mechanisms create social and cultural cohesion.

Socially relevant problems are rarely dealt with within the limits of one single system rationale. Instead, they go through interrelated procedures, either sequentially or in parallel. For example, the political system can decide on a specific goal or target by parliamentary vote (e.g., a limit on automobile emissions) and then leave it to the market to implement this decision (such as organizing an auction to sell emission rights to all potential emitters). Or a governmental decree is reviewed by an expert panel or a citizen advisory committee. Of particular interest are decision-making processes that combine the logic of two or more systems (Renn, 2008: 289). The settlement of conflicts with the method of mediation or negotiated rulemaking can, for example, be interpreted as a fusion of economic and social rationale. The cooperation between experts and political representatives in joint advisory committees (i.e., the experts provide background knowledge, while politicians highlight preferences for making the appropriate choices) represents a combination of knowledge-oriented elements and political governance. Classic hearings are combinations of expert knowledge, political resolutions, and the inclusion of citizens in this process.

3.3.2 Application to energy policies

Decisions about energy systems are routinely made within all four systems. Engineers and designers develop new technologies for more efficient energy conservation; managers may decide on a risky strategy to invest in new renewable energy technology; consumers may decide to invest in insulation material and a new solar heating system; and political bodies may impose a special carbon tax on all fossil fuels. Conflicts are likely to occur when new energy policies impose restrictions on one part of the population to provide more benefits to other parts or vice versa (Linnerooth-Bayer and Fitzgerald, 1996). In these cases, legitimate decision-making requires the proof that:

- alternative actions are less cost-effective (*efficiency*);
- the required course of action (or nonaction) would result in anticipated positive results (*effectiveness*);
- the actions are enhancing the functionality of the system even under stress conditions (*resilience*); and
- these actions are in line with public preferences and values and are accepted even by those who disagree with the decision (*reflection of social preferences and values, in particular fair distribution of benefits and costs*).

All four aspects feed into the overarching goal of legitimacy of energy policies. Legitimate decisions need compliance to due process rules and public acceptance on the basis of perceived effectiveness, efficiency, fairness, and resilience. When contemplating the acceptability of a proposed energy policy, one needs to be informed about the likely consequences of each decision option, the opportunity cost for choosing one option over the other, the sustainability of vital services to the communities, and the potential violations of interests and values connected to each decision option (McDaniels, 1996). This is basically true for any far-reaching political decision, but particularly to energy decisions that are under severe scrutiny by major actors in society (energy sector, NGOs).

3.4 Generic concepts for stakeholder and public involvement

3.4.1 Six concepts of inclusive governance

In Section 3.2 the two terms inclusion and closure were introduced. The insights from Section 3.3 suggest that that for complex energy decisions representatives of all four sectors of society need to be included in order to ensure that decisions are effective, efficient, resilient, and socially coherent. It seems also prudent to conclude that representatives of one sector should not be able to outvote the representatives of the others sectors since each contribution is needed for good decision-making. Maximizing efficiency on the expense of the other goals may compromise resilience and maximizing effectiveness may compromise social cohesion. However, the design of stakeholder involvement depends not only on the macro-functional view of what society needs for producing good decision but also need to include the expectations and perspectives of the actors within the involvement process. Stakeholders differ in their perspectives and expectations: for example, one might hope to get a representative judgment about choosing the most appropriate option for dealing with sustainable energy choices or another may expect the best available knowledge to be included in the decision-making process. Other may raise questions such as follows: Is the goal to reach a consensus or just a snap shot of diverse opinions? Should participants be educated before reaching a conclusion or should they rely on their given preferences to make public choices? Should everybody have an opportunity to shape the final product or only those with special knowledge about the subject or those who are most affected by the decision?

These questions cannot be answered without referring to the concepts or even philosophies underlying different perspectives on the role and function of participation in democratic societies. It all depends on which school of thought one implicitly or explicitly belongs. One can differentiate between six distinct prototypes of structuring processes that channel public input into public policy making. These prototypes can be labeled as functionalist, neoliberal, deliberative, anthropological, emancipatory, and postmodern (Renn, 2008: 294ff, 2014; US-National Academy, 2008; Renn and Schweizer, 2009). These six prototypes have to be looked upon as abstractions from real-world interaction to the extent that no participation process would be considered as belonging exclusively to one of these categories. Rather, they are ideal types in the Weberian sense (Weber, 1972). Originally, the perspectives on participation were derived from philosophical traditions. Today they serve as mental constructs of social reality, thus inspiring and informing different stakeholders about the role and function of involvement processes.

3.4.2 Functionalist concept

This approach to citizen participation draws on the functional school of social sciences and evolutionary concepts of social change. Functionalism is originally based on the works of Malinowski (1944) and Radcliffe-Brown (1935), the founding fathers of British and US functionalism. Functionalism conceptualizes society as a complex structure, recognizing essential functions for social survival either from an individual actor's perspective (Malinowski) or from society's point of view (Radcliffe-Brown). Each social action is assumed to be functional in assisting society's physical, social, and cultural sustainability (Hillmann, 1994: 252). As a later development primarily associated with Talcott Parsons and Robert K. Merton, structural functionalism presumes that a system has to meet functional imperatives (adaptation, goal attainment, integration, and latent patterns maintenance). These functions are performed by certain structures (Parsons, 1951; Merton, 1959).

In this sense, participatory exercises are necessary in order to meet complex functions of society that need input (knowledge and values) from different constituencies. The main objective is to avoid missing important information and perspectives, and to ensure that all knowledge camps are adequately represented. Participation is, therefore, seen as a process of getting all the problem-relevant knowledge and values incorporated within the decision-making process. The functionalist approach can be subdivided into two major functional goals: first, to collect all the necessary knowledge to solve a problem and, second, to avoid political paralysis by demonstrating openness to all stakeholders. Functionalist decision-making is clearly oriented toward goal achievement and synthesizing knowledge and values toward achieving a predefined goal. In terms of the basic functions of society as outlined above, the model is designed to improve and enhance the *effectiveness* of decision-making. It assumes that representation and inclusion of diversity will result in the improvement of environmental policy making with respect to the quality of the decisions made. Methods of participation suitable for this approach are expert Delphi methods, negotiated rulemaking, hearings, and citizen advisory committees (Webler et al., 1991; Gregory et al., 2001). These methods of participation are especially suited for the functional perspective because they emphasize the inclusion of various kinds of information for strategic planning.

3.4.3 Neoliberal concept

This approach to citizen participation draws on the philosophical heritage of liberalism and Scottish moral philosophy (Jaeger et al., 2001: 20ff). Neoliberalism conceptualizes social interaction as an exchange of resources. In this concept, deliberation is framed as a process of finding one or more decision option(s) that optimizes the payoffs to each participating stakeholder. The objective is to convert positions into statements of underlying interests (for a general overview, see Fisher and Ury, 1981; review of pros and cons in Jaeger et al., 2001: 243ff; Schweizer, 2008). The rational actor paradigm understands humans as resourceful and restricted individuals who have expectations, engage in evaluation, and maximize options.

Neoliberal decision-making consequently focuses on individual interests and preferences (Schweizer, 2008). It is assumed that people pursue their individual goals according to their available resources. The market is the place where these preferences can be converted into the appropriate actions under the condition that choices between different options are open to all individuals, and that the selection of options by each individual does not lead to negative impacts upon another individual's resources (absence of external effects). If all individuals have the resources to select options and all suppliers have the opportunity to offer options, the market guarantees optimal allocation and distribution of goods.

If, however, the aspired good requires collective action by many individuals, or if an individual good implies burdening society with external costs and benefits, the market mechanism will fail and public policies, including collectively binding norms and rules, are needed. These policies should reflect the preferences of all the individuals who are affected by the decision (Fisher and Ury, 1981). Since not all preferences are likely to represent identical goals and the means of achieving them, a negotiation process must be initiated that aims at reconciling conflicts between actors with divergent preferences. Within neoliberal theory, individual preferences are given so that conflicts can only be reconciled if, first, all of the preferences are known in the proportional distribution among all affected parties and, second, compensation strategies are available to recompense those who might risk utility losses when the most preferred option is taken. The two ideal outcomes of negotiation are, hence, to find a new win–win option that is in the interest of all or at least does not violate anybody's interest (Pareto optimal solution), or to find a compensation that the winner pays to the losers to the effect that both sides are at least equally satisfied with respect to the two choices: the situation before and after the compensation (Kaldor–Hicks solution). Deliberation helps to find either one of the two solutions and provides acceptable trade-offs between overprotection and underprotection for all participants.

Under these conditions, participation is required to generate a most truthful representation of public preferences within the affected population (Amy, 1983). The measurement of preferences is, however, linked to the idea that individuals should have the opportunity to obtain the best knowledge about the likely consequences of each decision option (concept of informed consent). Therefore public opinion polls are not sufficient to represent the public view on a specific public good or norm. Appropriate methods for revealing informed public preferences are referenda, focus groups, (internet) forums, roundtables, and multiple discussion circles (Ethridge, 1987; Dürrenberger et al., 1999). For the second objective to generate win–win solutions or acceptable compensation packages, negotiation, arbitration and, especially, mediation are seen as the best instrumental choices (Amy, 1987; Baughman, 1995). These methods correspond with the neoliberal emphasis on bargaining power and balancing individual interests. The main contribution of neoliberal participation models is to be more *efficient* and, to a lesser degree, to be more *reflective of social values and concerns* as they are distributed among the given population.

3.4.4 Deliberative concept

Deliberative citizen participation is mainly influenced by Habermasian discourse theory (Habermas, 1984; Webler, 1995; Renn et al., 1993: 48–57). Discourse theory and discourse ethics advocate more inclusiveness for legitimate and sustainable political decision-making. Modern societies are characterized by a plurality of values and worldviews. According to Habermas (1996: 20), conventional politics and political decision-making cannot adequately deal with this heterogeneity. Modern societies lack moral cohesion that could guide political decision-making. Although mutually binding norms and values are nonexistent at the surface, people can allude to their shared reason and experience as human beings. Here, the joined heritage of Habermasian deliberation and "communitarism" becomes obvious (Benhabib, 1992). Consequently, political decision-making has to find mechanisms that could serve as guidance instruments by enabling citizens to engage in joint rational decision-making.

Habermasian discourse ethics offers a solution to this dilemma. In discourse ethics, only those political and judicial decisions may claim to be legitimate that may find the consent of all affected parties. Consent does not merely mean agreement but is a product of a rule-based discursive opinion formation and decision-making process (Habermas, 1992: 169). This process can best be described as a competition of arguments. As a result, the procedure of decision-making decides on its legitimacy in the broad sense explained above. Habermas claims that in communication, people always make one or more factual, normative, or subjective knowledge claims (in his words: speech acts). These claims allude to the objective world of factual evidence, the normative world of values, moral orientations and worldviews, and the subjective world of individual experience. The basic premise of the theory of communicative action is that people are capable of coming to a rationally motivated agreement (i.e., agreements free of coercion of any kind) if they are provided with the optimal discourse setting.

Thus factual, normative, and expressive knowledge claims are settled by alluding to the common rationality of communicative action provided by an appropriate organizational discourse structure. Of course, no real-world discourse can reach the prerequisites of the ideal speech situation (Webler, 1995); yet, practical discourse can aspire to this goal. Discursive decision-making is therefore oriented toward the common good and seeks the rational competition of arguments. It looks for diversity in participants and perspectives in the sense that all potentially affected parties should be able to agree with its outcome. All relevant arguments need to be included in the deliberation regardless of the extent of their representation within the population. The objective here is to find the best possible consensus among moral agents (not just utility maximizers) about shared meaning of actions based on the knowledge about consequences and an agreement on basic human values and moral standards (Webler, 1995, 1999). The results of discursive decision-making then draw their legitimization from the procedural arrangements of the discourse. Participation methods aim at facilitating mutual understanding and transparent decision-making, thus adding legitimacy to the whole process of policy making. The best-suited instruments refer to citizen forums, multiple stakeholder conferences, and consensus-oriented meetings (Dienel, 1989; Rowe and Frewer, 2000; Rowe et al., 2004). The main contribution of deliberative models to society is to enhance *resilience and legitimacy* and to reflect *social and cultural values in collective decision-making*.

3.4.5 Anthropological concept

Anthropological citizen participation is mainly influenced by pragmatic Anglo-Saxon philosophy. It is based on the belief that common sense is the best judge for reconciling competing knowledge and value claims. Pragmatism was mainly influenced by the works of Pierce (1867) and Dewey (1940) (review in Hammer, 2003).

For participatory decision-making, this approach has far-reaching consequences. The moral value of policy options can be judged according to their consequences. Furthermore, citizens are equally capable of moral judgment without relying on more than their mind and experience. When organizing discourses of this kind, however, there is a need for independence, meaning that the jury has to be disinterested in the topic and there should be some consideration of basic diversity in participants (such as gender, age, and class). The goals of decision-making inspired by the anthropological perspective are the involvement of the "model" citizen and the implementation of an independent jury system consisting of noninterested laypersons who are capable of employing their common sense for deciding on conflicting interests (Stewart et al., 1994; Sclove, 1995). Participatory methods granting this kind of common-sense judgment are consensus conferencing, citizen juries, and planning cells (Crosby et al., 1986; Dienel, 1989; Abels, 2007). The group of selected individuals can be small in size. Most methods do not require more than 12–25 participants to accomplish valid results (Stewart et al., 1994). Within that small number, there should be a quota representation of the entire population, thus including the general perspectives of all citizens. The main focus of the anthropological model is to reflect *social values and concerns* in public policy making, in particular *fairness*.

3.4.6 Emancipatory concept

The basic ideas of emancipatory participation are derived from a Marxist or neo-Marxist social perspective (Ethridge, 1987; Jaeger et al., 2001: 232ff). The goal of inclusion is to ensure that the less privileged groups of society are given the opportunity to have their voices heard and that participation provides the means to empower them to become more politically active (Fischer, 2005). In the long run, participation is seen as a catalyst for an evolutionary, or even revolutionary, change of power structures in capitalist societies (Fung and Wright, 2001).

The main motive for participation is the revelation of hidden power structures in society. This motive is shared by the postmodern school described later. Yet, the main emphasis in the emancipatory school is the empowerment of the oppressed classes to, first, acknowledge their objective situation and then become aware of their own resources to change the negative situation in which they live, develop additional skills and means to fight these unjust structures, and, lastly, be prepared to continue this fight even after the participatory exercise is completed. The thrust is the awakening of individuals and groups to make them more politically active and empowered (Skillington, 1997). The goal of empowerment has spread out to left or liberals positions in the politics. Crucial application areas are the pursuit of environmental justice, community development, access to basic services to the poor, and enhancing rural development (Fung and Wright, 2001). Several analysts have linked empowerment and the mobilization of resources to the theory of social capital (Larsen et al., 2004).

Methods within the emancipatory concept include activist-driven public meetings, tribunals, science shops, and community solidarity committees (McCormick, 2007). The main emphasis is on making sure that the powerless are heard and then empowered to fight for their own interests and values. Although the focus of this concept is on transforming society; it does add to a more balanced *reflection of social and cultural values* in the policy-making process. Fairness in this perspective entails more equality in society and even access to resources and societal opportunities.

3.4.7 Postmodern concept

This approach to citizen participation is based on Michel Foucault's theory of discourse analysis. Discourse analysis rests on the three basic concepts of knowledge, power, and ethics. Foucault is interested in the constitution of knowledge. He assumes that knowledge formation is a result of social interaction and cultural settings. Truth then depends upon historically and socially contingent conditions (Foucault, 2003) and is not independent of the subject that makes truth claims. In this sense, postmodern decision-making aims at revealing the hidden power and knowledge structures of society, thus demonstrating the relativity of knowledge and values (Fischer, 2005: 25). Reaching a consensual conclusion is neither necessary nor desirable. In its deconstructivist version, deliberation serves as an important social experience that any universal claim to truth and ethical values is bound to fail and that validity can only be claimed for one's own subjective domain of beliefs and convictions. These may be shared with others but there are no overarching arguments or supra-rationality for justifying universally valid claims. In its constructive version, deliberation leads to the enlightenment of decision-makers and participants as to be informed about the plurality of positions and arguments in a given debate (Jaeger et al., 2001: 221ff). Far from resolving or even reconciling conflicts, deliberation, according to this viewpoint, has the potential to decrease the pressure of conflict, to provide a platform for making and challenging claims, and to assist policy makers in coping with diversity (Luhmann, 1989). Deliberations help reframe the decision context, make policy makers aware of public demands, and enhance legitimacy of collective decisions through reliance on formal procedures (Skillington, 1997). The process of talking to each other, exchanging arguments, and widening one's horizon is what deliberation can accomplish. It is an experience of mutual learning without a substantive reliance on a superior logic or rationality.

Participatory decision-making seeks to include dissenting views and social minorities, thus illustrating the relativity of knowledge and power. Appropriate participatory methods include framing workshops, discussion groups, internet chat rooms, and open forums because they do not set rigid frames for decision-making (Stirling, 2004). Rather, they provide insight into stakeholder interests, knowledge bases, and power structures. Accordingly, the main function of postmodern discourse is to enlighten the policy process by illustrating the *diversity of factual claims, opinions, and values*.

3.4.8 Implications of the different concepts for practical discourse

This review of different perspectives for stakeholder involvement in collective decision-making is more than an academic exercise. Organizers, participants, observers, and the addressees of public participation are implicitly or explicitly guided by these concepts (Webler, 2011). Often, conflicts about the best structure of a participatory process arise from overt or latent adherence to one or another concept. Advocates of neoliberal concepts stress the need for proportional representation (i.e., representativeness) of participatory bodies, while advocates of deliberative concepts are satisfied with a diversity of viewpoints. For advocates of the anthropological model, representativeness plays hardly any role as long as common sense is ensured. Models driven by emancipatory concepts will judge the quality of participation by the degree to which underprivileged groups have gained more access to power, whereas functionalist models will judge the quality of the process by the quality of the outputs compared to either technocratic or decisionistic (synthesis of knowledge from experts and values from politicians) decision-making models. While neoliberal concepts will take public preferences as a given prerogative to participatory decision-making, deliberative models are meant to influence preferences and change them through the process. Table 3.1 provides an overview of the six models, their main rationale, and some of the instruments, which can be associated with them.

The diversity of concepts and background philosophies is one of the reasons why participatory processes are so difficult to evaluate in terms of overarching evaluative criteria (Rowe and Frewer, 2000, Rowe et al., 2004; Renn, 2008: 320ff). Although some of these models can be combined and integrated, there are at least differences in priorities. It is obvious that within the functionalist school, the main evaluation criterion is the quality of the output, whereas the models inspired by postmodernism and emancipatory schools are not interested in output but, rather, in the changes that were induced in the minds of the participating people (raising awareness and emancipation).

TABLE 3.1 The six concepts of stakeholder involvement and their salient features.

Concept	Main objective	Rationale	Models and instruments
Functionalist	To improve the quality of decision output	Representation of all knowledge carriers; integration of systematic, experiential, and local knowledge	Delphi method, workshops, hearing, inquiries, citizen advisory committees
Neoliberal	To represent all values and preferences in proportion to their share in the affected population	Informed consent of the affected population; Pareto-rationality plus Kaldor–Hicks methods (win–win solutions as described in Chapter 2: History of the energy transition in Germany: from the 1950s to 2019)	Referendum, focus groups, internet-participation, negotiated rulemaking, mediation, etc.
Deliberative	To debate the criteria of truth, normative validity, and truthfulness	Inclusion of relevant arguments, reaching consensus through argumentation	Discourse-oriented models, citizen forums, and deliberative juries
Anthropological	To engage in common sense as the ultimate arbiter in disputes (the jury model)	Inclusion of noninterested laypersons representing basic social categories (e.g., gender, income, and locality)	Consensus conference, citizen juries, and planning cells
Emancipatory	To empower less privileged groups and individuals	Strengthening the resources of those who suffer most from environmental degradation	Action group initiatives, town meetings, community development groups, tribunals, and science shops
Postmodern	To demonstrate variability, plurality, and legitimacy of dissent	Acknowledgment of plural rationalities; no closure necessary; mutually acceptable arrangements are sufficient	Open forums, open space conferences, and panel discussions

Given this mix of models driven by different concepts, many participation analysts and practitioners have advocated hybrid models that combine elements of different models. Endeavors to combine the neoliberal with the deliberative concept include the deliberative polling method, which has been widely used in several areas of environmental policy making (Ackerman and Fishkin, 2004). More complex hybrid models try to include even more than two concepts, such as the cooperative discourse model (Renn, 1999). For dealing with risk problems, it seems particularly helpful to combine the functionalist with the deliberative process. Such a combination has been accomplished with the concept of analytic–deliberative discourse developed and advocated by the US National Academy of Sciences (Stern and Fineberg, 1996; US National Science Council, 2008). The following section describes this approach in more detail and demonstrates why the analytic–deliberative approach is particularly suited for energy policy making.

For making prudent energy choices, there has been a strong preference for the functionalist and neoliberal view of participation. Many policy makers in the field of energy have been, and still are, primarily interested in input from the relevant stakeholders in order to improve the quality of the decisions and to make sure that conflicting values could be resolved in proportion to the representation of the people affected by the decision (Fiorino, 1990). More lately, there has been a shift toward deliberative and emancipatory forms of participation (Nanz and Leggewie, 2019). The discussion on environmental justice, as well as on social capital, has served as a catalyst for these more intense forms of argument-based participation (Dryzek, 1994). In parallel the anthropological concept has inspired many organizers of participation to model participation in accordance with the well-established jury format of the US judicial system (Crosby et al., 1986).

In our view a combination of the functional and deliberative concepts is most suitable for dealing with complex energy problems. In order to justify this opinion, it is necessary to discuss some specific features of the energy debate that highlight the need for functional problem-solving and intense value deliberation.

3.5 The analytic–deliberative approach to stakeholder involvement

3.5.1 Combing analysis and deliberation

One particularly promising suggestion for combining functional and deliberative decision-making is the model of analytic–deliberative discourse (Stern and Fineberg, 1996; Tuler and Webler, 1999; Renn, 2014). Such a process is designed to provide a synthesis of scientific expertise, a common interpretation of the analyzed relationships and a balancing of pros and cons for making a final decision based on insights and values. Analysis in this context means the use of systematic, rigorous, and replicable methods of formulating and evaluating knowledge claims (Stern and Fineberg, 1996: 98; see also Tuler and Webler, 1999: 67). These knowledge claims are normally produced by scientists (natural, engineering, and social sciences, as well as the humanities). In many instances, relevant knowledge also comes from stakeholders or members of the affected public (Horlick-Jones et al., 2007). Deliberation highlights the style and nature of problem-solving through communication and collective consideration of relevant issues (Chambers, 2003; original idea of discursive deliberation from Habermas, 1970). It combines different forms of argumentation and communication, such as exchanging observations and viewpoints, weighing and balancing arguments, offering reflections and associations and putting facts into a contextual perspective. The term deliberation implies equality among the participants, the need to justify and argue for all types of (truth) claims and an orientation toward mutual understanding and learning (Dryzek, 1994; Renn, 2008). The following subsections explain the two components of an analytic–deliberative process: analysis and deliberation (see similar sections in Rosa et al., 2014: 178ff).

3.5.2 The first component: the role of scientific analysis

The first element of analytic–deliberative processes refers to the inclusion of systematic and reproducible knowledge for making prudent energy decisions. There is little debate in the literature about the inclusion of external expertise being essential as a major resource for obtaining and utilizing systematic and experiential knowledge (Horlick-Jones et al., 2007). The following four points show the importance of knowledge for risk management but also make clear that choosing the right management options requires more than looking at the scientific evidence alone.

First, scientific input is essential for designing energy policies. The degree to which the results of scientific inquiry are taken as ultimate evidence to judge the appropriateness and validity of competing knowledge claims may be contested and should, therefore, be one of the discussion points during deliberation. The answer to this question may depend on context and the maturity of scientific knowledge in the respective risk area. [A similar assessment is provided in Horlick-Jones et al. (2007).] For example, if the issue is the effect of a specific emission such as nitrogen oxide (from Diesel engines) on human health, anecdotal evidence may serve as a heuristic tool for further inquiry, but there is hardly any reason to replace toxicological and epidemiological investigations with intuitions from the general public. If the issue is the siting of a wind park, anecdotal and local knowledge about sensitive ecosystems or traffic flows may be more relevant than systematic knowledge about these impacts in general (Renn, 2010).

Second, the resolution of competing claims of scientific knowledge should be governed by the established rules within the respective discipline. These rules may not be perfect and may even be contested within the community. Yet they are usually superior to alternatives (Shrader-Frechette, 1991: 190ff).

Third, many problems and decision options require systematic knowledge that is unavailable, still in its infancy, or in an intermediary status. Analytic procedures are demanded as a means of assessing the relative validity of each of the intermediary knowledge claims, show their underlying assumptions and problems, and demarcate the limits of reasonable knowledge (i.e., identify the range of those claims that are still compatible with the state-of-the-art in this knowledge domain) (SAPEA, 2019).

Fourth, knowledge claims can be *systematic* and scientific, as well as *experiential* (based on long-term familiarity with energy issue in question); *local* (referring to one's own experience with the local conditions surrounding wind or solar parks); or derived from *folklore wisdom*, including common sense (Renn, 2010). All of these forms of knowledge have a legitimate place in analytic–deliberative processes. How they are used depends on the context and the type of knowledge required for the issue under question. For example, if a waste-to-energy conversion facility is going to be sited, systematic knowledge is needed to understand the dose–response relationships between the flue gas and potential health or environmental damage; experiential knowledge is needed for assessing the reliability of the control technology or the sincerity of the plant's operator to implement all of the required control facilities; local knowledge may be helpful to consider special pathways of diffusion of gases or special routes of exposure; and common sense or folklore wisdom may assist decision-makers in ordering and prioritizing multiple knowledge claims.

3.5.3 The second component: deliberation

The term *deliberation* refers to the style and procedure of decision-making without specifying the participants who are invited to deliberate (Chambers, 2003; Stern and Fineberg, 1996). Deliberation is foremost a style of exchanging arguments and coming to an agreement on the validity of statements and inferences. Using a deliberative format does not necessarily include the demand for stakeholder or public involvement. Deliberation can be organized in closed circles (such as conferences of Catholic bishops, where the term has, indeed, been used since the Council of Nicaea), as well as in public forums. Following our arguments, however, complex risk decisions require contributions from scientists, policy makers, stakeholders, and affected publics, and a procedure is required that guarantees both the inclusion of different constituencies outside the risk-management institutions and the assurance of a deliberative style within the process of decision-making. We use the term *deliberative democracy* to refer to the combination of deliberation and third-party involvement (see also Warren, 2002).

In terms of making prudent energy decisions, deliberation is required for a variety of reasons. First, it can produce common understanding of the issues or the problems based on the joint learning experience of the participants with regard to systematic and anecdotal knowledge. Furthermore, it may produce a common understanding of each party's position and argumentation (rationale of arguing) and thus assist in a mental reconstruction of each actor's argumentation (Habermas, 1970; Warren, 1993). The main drive in gaining mutual understanding is empathy. Habermas's theory of communicative action (HCAT) provides further insights about how to mobilize empathy and how to use the mechanisms of empathy and normative reasoning to explore and generate common moral grounds (Webler, 1995).

Second, deliberation can produce new options for action and solutions to a problem. This creative process can be mobilized either by finding win–win solutions or by discovering identical moral grounds on which new options can grow (Fisher and Ury, 1981; Webler, 1999). It has the potential to show and document the full scope of ambiguity associated with risk problems and helps to make a society aware of the options, interpretations, and potential actions connected with the issue under investigation (De Marchi and Ravetz, 1999). Each position within a deliberative discourse can survive the cross fire of arguments and counterarguments only if it demonstrates internal consistency, compatibility with the legitimate range of knowledge claims, and correspondence with the widely accepted norms and values of society. Deliberation clarifies the problem, makes people aware of framing effects, and determines the limits of what could be called reasonable within the plurality of interpretations (Skillington, 1997).

Third, deliberation can produce common agreements. The minimal agreement may be a consensus about dissent (Renn et al., 1993: 64). If all arguments are exchanged, participants know why they disagree. They may not be convinced that the arguments of the other side are true or morally strong enough to change their position, but they may understand the reasons that the opponents came to their conclusions. At the end the deliberative process produces several consistent and—in their own domain—optimized positions that can be offered as package options to legal decision-makers or the public. Once these options have been subjected to public discourse and debate, political bodies such as agencies or parliaments can make the final selection in accordance with legitimate rules and institutional arrangements, such as a majority vote or executive order. Final selections also can be performed by popular vote or referendum. In addition, deliberation creates "second-order" effects on individuals and society by providing insights into the fabric of political processes and by creating confidence in one's own agency to become an active participant in the political arena (Dryzek, 1994). By participating, individuals can enhance their capacity to raise their voice in future issues and become empowered to play their role as active citizens in the various political arenas.

Fourth, deliberation may result in consensus. Often, deliberative processes are used synonymously with consensus-seeking activities (Coglianese, 1997). This is a major misunderstanding. Consensus is a possible outcome of deliberation but not a mandatory requirement (cf. van den Hove, 2007). If participants find a new option that they all value more than the option they preferred when entering the deliberation, a "true" consensus is reached (Habermas, 1970). But it is clear that finding such a consensus is the exception rather than the rule. Consensus is either based on a win–win solution or a solution that serves the "common good" and each participant's interests and values better than any other solution (see, e.g., Webler et al., 2001). The requirement of a *tolerated* consensus is less stringent. Such a consensus rests on the recognition that the selected decision option might serve the common good best but at the expense of some interest violations or additional costs. Here participants might agree to a common decision outcome while holding their nose. In this situation, people who might be worse off than before, but who recognize the moral superiority of the solution, can abstain from using their power of veto without approving the solution. In our own empirical work, deliberation often has led to a tolerated consensus solution, particularly in siting conflicts (one example is Schneider et al., 1998). Consensus and tolerated consensus should be distinguished from *compromise*. A compromise is a product of bargaining, with each side gradually reducing its claim to the opposing party until they reach an agreement (Raiffa, 1994). All parties involved would rather choose the option they preferred before starting deliberations, but because they cannot find a win–win situation or a morally superior alternative, they look for a solution that they can "live with," well aware of the fact that it is the second- or third-best solution for them. Compromising on an issue relies on full representation of all vested interests (Box 3.1).

BOX 3.1

Example of an analytic–deliberative process on climate change polices

In 2013 the Ministry of Environment and Energy of the German State of Baden Württemberg asked the nonprofit company Dialogik (codirected by coauthor Ortwin Renn) to initiate a major public involvement program for an integrated State Policy on Climate Change. The ministry had commissioned a scientific consultant to come up with more than 100 possible policy suggestions from which the stakeholders were asked to select the most appropriate measures and to prioritize them for the next 10 years (Renn, 2014).

Dialogik established a total of 16 roundtables, 8 with organized stakeholder groups, four with randomly selected citizens from the State and two with volunteers from an internet search. The roundtables discussed different topical areas such as sustainable mobility, replacement of fossil fuel, energy efficiency in industry, energy efficiency in households, etc. However, out of the 8 stakeholders tables, two pursued the same topic, each independent of each other. Each stakeholder table had around 16 members, equally distributed among representatives of industry, science, nongovernmental organizations, and pubic administration. In addition, we took random samples of citizens for four roundtables devoted to issues that demanded behavioral adaptations by everyday citizens. Each citizen group had between 20 and 30 participants. Again, identical topics were discussed in two of the four citizen tables. Finally, Dialogik organized an internet forum that attracted more than 12,000 active participants from which it asked for volunteers to recruit for one or more extra tables. There were enough volunteers for establishing two extra roundtables. Both dealt with the issue of improving energy efficiency in households.

Each table produced a list of measures in three categories: high priority, rejected, and tolerable. To facilitate the selection and evaluation process, Dialogik had participants rate each measure on four criteria: effectiveness, efficiency (cost-effectiveness), fairness, and robustness (the term resilience was not familiar to many of the participants). In addition, the group members could suggest measures of their own or modify those that had been given to them. The moderators of each group were instructed to accomplish a consensus, but in three of the 16 groups, a majority and minority list was produced. Then the tables selected two spokespersons, who met with the other spokespersons of the groups with the same topic in order to discuss the list and make the necessary adjustments. The final round was a meeting with the spokespersons of all 20 groups. Although there was no consensus on all the measures, the Dialogik moderator was able to elicit 18 high priority measures that were confirmed and approved by everyone and 34 measures that all participants were willing to tolerate.

The State Ministry honored its promise and reviewed all the recommendations and comments in great detail. In total the citizens panels produced 272, the stakeholder roundtables 323, and the joint meetings of the spokespersons for each topic additional 145 recommendations and comments for the ministry. As a result of this productive involvement process, the administrators at the ministry took 6 months to prepare a

BOX 3.1 (*cont'd*)

report specifying which measures they will adopt, modify, or reject (and the reasons why). The whole process was accompanied by an external evaluation, which demonstrated high satisfaction with the process by all participants and a particular appreciation of the inclusion of all major stakeholders. In 2019 this participatory process won the first price in a national competition for the best citizen participation project in the domain of policy planning.

3.5.4 Empirical evidence: what works?

The discussion so far has focused on the potential of analytic–deliberative processes, their advantages and disadvantages, and the interface between the analytic and the deliberative process. This section deals with the internal structure of deliberation. There is a need for an internal structure that facilitates common understanding, rational problem-solving, and fair and balanced treatment of arguments. The success or failure of a discourse depends on many factors. Among the most influential are the following (Renn, 2008: 318ff; Renn et al., 1993: 57ff):

- *A clear mandate for the participants of the deliberation.* Models of deliberative democracy require a clear and unambiguous mandate of what the deliberation process should produce or deliver. Since deliberations are most often informal instruments, there should be a clear understanding that the results of such a process cannot claim any legally binding validity (unless the process is part of a legal process, such as arbitration). All of the participants, however, should begin the process with a clear statement that specifies their obligations or promises of voluntary compliance once an agreement has been reached.
- *Openness regarding results.* A deliberative process will never accomplish its goal if the decision already has been made (officially or secretly) and the purpose of the communication effort is to "sell" this decision to the other parties. Individuals have a good sense of whether a decision maker is really interested in their point of view or if the process is meant to pacify potential protesters.
- *A clear understanding of the options and permissible outcomes of such a process.* The world cannot be reinvented by a single involvement process nor can historically made decisions be deliberately reversed. All participants should be clearly informed of the ranges and limits of the decision options that are open for discussion and implementation. If, for example, the technology is already in existence, the discourse can focus only on issues such as emission control, monitoring, emergency management, or compensation. But the range of permissible options should be wide enough to provide a real choice situation to the participants.
- *A predefined timetable.* It is necessary to allocate sufficient time for all the deliberations, but a clear schedule, including deadlines, is required to make the process effective and product oriented.

- *A well-developed methodology for eliciting values, preferences, and priorities.* The need for efficiency in risk governance demands a logically sound and economical way to summarize individual preferences and integrate them within a group decision (agreement on dissent, majority and minority positions, tolerated consensus, true consensus, or compromise). Formal procedures, such as multiobjective or multiattribute utility analysis, could serve as tools for reaching agreements (von Winterfeldt and Edwards, 1986; Hostmann et al., 2005). We have used multiattribute utility procedures in most of our deliberative processes (Renn, 2008).
- *Equal position of all parties.* A deliberative process needs a public sphere with a climate of a "powerless" environment. This does not mean that every party has the same right to intervene or claim a legal obligation to be involved in the political decision-making process. However, the internal discourse rules have to be strictly egalitarian; every participant must have the same status in the group and the same rights to speak, make proposals, or evaluate options. Two requirements must be met. First, the decision about the procedure and the agenda must rely on consensus; every party needs to agree. Second, the rules adopted for the discourse need to be binding for all members, and no party can be allowed to claim any privileged status or decision power. The external validity of the discourse results, however, is subject to all legal and political rules that are in effect for the topic in question.
- *Neutrality of the facilitator.* The person who facilitates such a process should be neutral in his or her position on the risk issue and respected and authorized by all participants. Any attempt to restrict the maneuverability of the facilitator, moderator, or mediator should be strictly avoided.
- *A mutual understanding of how the results of the process will be integrated within the decision-making process of the regulatory agency.* As a predecisional tool, the recommendations cannot, in most cases, serve as binding decisions. They should instead be regarded as a consultative report similar to the technical recommendations provided by scientific consultants to the legitimate public authorities (National Research Council, 2008). Official decision-makers must acknowledge and process the reports by the deliberative bodies, but they are not obliged to follow their advice. However, the process will fail its purpose if deviations from the recommendations are neither explained nor justified to the panelists.

A second set of internal requirements about the expected behavior of the participants is necessary to facilitate agreement or, at least, a productive discussion. Among these, the requirements are the following:

- *Willingness to learn.* All parties must be ready to learn from each other. This does not necessarily imply that they have to be willing to change their preferences or attitudes. Conflicts can be reconciled on the basis that parties accept others' positions as a legitimate claim without giving up their own point of view. Learning in this sense entails:
  - recognition of various forms of rationality in decision-making
  - recognition of different forms of knowledge, whether it is systematic experiential, local, or indigenous (Renn, 2010); and

- willingness to subject oneself to the rules of argumentative disputes (i.e., provide factual evidence for claims), obey the rules of logic for drawing inferences, and disclose one's own values and preferences vis-à-vis potential outcomes of decision options.
- *Resolution of allegedly irrational responses.* Reflective and participatory discourses frequently demonstrate a conflict between two contrasting modes of evidence. The public refers to anecdotal and personal evidence, mixed with emotional reactions, whereas the professionals employ their systematic and generalized evidence based on abstract knowledge. A dialogue between these two modes of collecting evidence is rarely accomplished, because experts regard the personal evidence as a typical response of irrationality. The public representatives often perceive the experts as uncompassionate technocrats who know the statistics but could not care less about a single life lost. This conflict can be resolved only if both parties are willing to accept the rationale of the other party's position and to understand, and maybe even empathize with, other party's point of view. If over the duration of discourse, some familiarity with the process and mutual trust among the participants have been established, role playing can facilitate that understanding. Resolving alleged irrationalities means discovering the hidden rationality in the argument of the other party.
- *Nonmoralization of positions and parties.* The individuals involved in a deliberative process should agree in advance to refrain from moralizing. Moral judgments on positions or persons impede compromise. As soon as parties start to moralize, they cannot make trade-offs between their allegedly moral position and the other parties' "immoral" position without losing face (Renn, 2004). A second undesired result of moralizing is the violation of the equality principle. Nobody will assign equal status to a party who is allegedly morally inferior. Finally, moralizing masks deficits of knowledge and arguments. Even if somebody knows nothing about a subject or has only weak arguments to support his or her position, assigning blame to other actors and making the issue moral one can help win points. The absence of moralizing does not mean refraining from using ethical arguments, such as "This solution does not seem fair to future generations" or "We should conserve this ecosystem for its own sake." Ethical arguments are essential for resolving risk disputes.

The website of the International Risk Governance Council provides practical guidance on how to implement and control these major rules for effective and fair stakeholder representation in decision-making processes (www.irgc.org/risk-governance/stakeholder-engagement-guide/). Most of the manuals listed on the website offer checklists or a sequence to show how to accomplish an effective, efficient, and fair process. Although each manual has its own approach to stakeholder involvement, they all agree on the following four major aspects that any program in this field needs to pay attention to:

- Start with a critical review of your own performance.
- Design an integrative analytic and deliberative program that ensures a continuous effort to communicate with and between experts, policy makers, and the most important stakeholders.
- Tailor the involvement process according to the needs of the targeted audience and the needs of the addressee(s) of the recommendations.

- Adjust and modify the stakeholder involvement process in an organized effort to collect feedback and to determine any changes in values and preferences. An empirical evaluation of what the process has accomplished is necessary in order to improve the process over time.

Stakeholder involvement programs pursue different purposes and objectives. All manuals agree, however, that—regardless of purpose or goal—each program must establish a common denominator on which the involvement can proceed and develop (Webler et al., 2001). This requires a good understanding of the stakeholders' specific needs for being engaged in an ongoing dialogue and participation program with respect to the issue under dispute. Since these issues can vary greatly, depending on the context and the history of the debate, a good understanding of all the circumstances—achieved prior to an engagement with the participants of the program—is a fundamental requirement in any effective involvement program.

3.6 Special requirements for participatory practices in the field of for energy policies

3.6.1 Linear, complex, uncertain, and ambiguous conditions

So far we discussed choice situations that require knowledge about the likely consequences of each decision option, the opportunity cost for choosing one option over the other, and the potential violations of interests and values connected to each decision option (McDaniels, 1996). This is basically true for any far-reaching political decision. Decision-making on complex energy issues, however, includes three additional elements that make decision-making more difficult: complexity, uncertainty, and ambiguity (Klinke and Renn, 2019; IRGC (International Risk Governance Council), 2017). What do we mean with these terms (IRGC (International Risk Governance Council), 2005; Renn, 2008: 177ff)?

- *Complexity*. Complexity is introduced when the causal relationship forms a multifaceted web of causal relationships, where many intervening factors may interact to affect the outcome of an event or an activity (WBGU, 2000: 194ff). Complexity requires sophisticated modeling, which often defies common sense or intuitive reasoning. Yet, if resolved, it produces a high degree of confidence in the results. Climate modeling may be taken as a prominent example here.
- *Uncertainty*. The less well known and understood this causal web is the more uncertainty is introduced into the system. Uncertainty reduces the strength of confidence in the estimated cause–effect chain (van Asselt, 2000). Far-reaching energy decisions must consider more carefully the uncertainties which characterize both the benefits and the risks. This particularly true for integrated energy systems linking power, heat, and mobility services.
- *Ambiguity*. Ambiguity arises when differences exist in how individual actors or stakeholders value some input or outcome of the system (IRGC (International Risk Governance Council), 2005: 30). It is based on the question of what our knowledge about risks mean for understanding the effects of the risk agent on human health and

the environment (interpretative ambiguity), and what kind of decisions or actions are justified once the risks and uncertainties are characterized (normative ambiguity). In energy policy making ambiguity plays an important role because plural knowledge and value input are difficult to reconcile and overarching arguments which might lead to a consensus are hard to find or to be approved by all parties (Jasonoff, 1998). A good example is the use of carbon sequestration techniques that may be framed as instrument for continuing the use of fossil fuels or as a means to avoid the most drastic implications of climate change in a transitional period.

Depending on the composition of complexity, uncertainty, and ambiguity, different levels of public and stakeholder participation seem appropriate to guarantee the quality of the process, if time and effort of the participating groups are regarded as spare resources. Four types of *"discourses,"* describing the extent of participation, have been suggested (Klinke and Renn, 2012).

In the case of simple *linear issues* with obvious consequences, low remaining uncertainties and no controversial values implied, such as investing in more home insulation, improving efficiency of technologies, or optimizing net structures, it seems not necessary and even inefficient to involve all potentially affected parties to the process of decision-making. An "instrumental discourse" is proposed to be the adequate strategy to deal with these situations. In this first type of discourse, policy makers, directly affected groups (like product or activity providers and involved individuals) and enforcement personnel are the relevant actors. It can be expected that the interest of the public into the regulation of these simple energy measures is very low. However, regular monitoring of the outcomes is important, as the issue might turn out to be more complex, uncertain, or ambiguous than characterized by the original assessment.

In the case of *complex problems* another discourse is needed. An example for complexity based problems is the so-called lock-in-effects of previous decisions that determine the combination of supply and demand. Many energy technologies are designed to be complementary to a nested structure of services, auxiliary technologies, and market designs. As complexity is a problem of insufficient understanding of the system as such, it is the more important to produce transparency over what is known and what is unknown about the system at hand—This "epistemic discourse" aims at bringing together the knowledge from the agency staff of different scientific disciplines and other experts from academia, government, industry, or civil society. The principle of inclusion is bringing new or additional knowledge into the process and aims at resolving cognitive conflicts.

In the case of energy problems due to large *uncertainties*, the challenges are even higher. The problem here is the following: how can one judge the severity of a situation when it is extremely difficult to predict the occurrence of events and/or their consequences? This dilemma concerns the characterization of the likely impacts as well as the evaluation and the design of options for dealing with these impacts. For example, a digital virtual system of pricing electricity (smart grid) is a highly complex but also uncertain intervention into the energy system where the impacts are only partially known and calculable. In this case, it is no longer sufficient to include experts into the discourse, but policy makers and the main stakeholders should additionally be included, to find consensus on how far one should go in this direction and who is bearing the costs and who is reaping the benefits.

This type is called "reflective discourse," because it is based on a collective reflection about balancing the possibilities of being overcautious or not courageous and innovative enough. For this type of discourse, other forms of participation are required (cf. Rowe and Frewer, 2000).

If the energy problems are due to *high ambiguity*, the most inclusive strategy is required, as not only the directly affected groups have something to contribute to the debate, but also the indirectly affected groups. If, for example, decisions have to be taken concerning the phase-out of coal energy in a country with many mines and coal fired power plants, this decision touches upon principal values and ethical questions, in particular, the issues of fairness. A "participatory discourse" has to be organized, where competing arguments, beliefs, and values can be openly discussed. This discourse affects the very early step of framing and designing new options. The aim of this type of discourse is to resolve conflicting expectations through identifying common values, defining options to allow people to live their own visions of a "good life," to find equitable and just distributions rules deciding benefits and costs, and to activate institutional means for reaching common welfare so that all can profit from the collective benefits.

In this typology of discourses, it is presupposed, that the categorization of energy issues into simple, complex, uncertain, and ambiguous is uncontested. But, very often, this turns out to be complicated. Who decides whether a special energy issue can be categorized as simple, complex, uncertain, or ambiguous? To resolve this question a *metadiscourse* is needed, where the decision is taken, where a specific problem is located and in consequence, to which route it is allocated. This discourse is called "design discourse," and is meant to provide stakeholder involvement at this more general level. Allocating each energy issue to one of the four routes has to be done before assessment starts, but as knowledge and information may change during the governance process, it may be necessary to reorder the allocation later in the implementation phase. A means to carry out this task can be a screening board that should consist of representatives of science, civil society, business, and political institutions.

Fig. 3.2 provides an overview of the described discourses depending on the three problem characteristics and the actors included into these discourses. In addition, it sets out the type of conflict produced through the plurality of knowledge and values and the required remedy to deal with the corresponding risk.

Of course, this scheme is a simplification of all the potential energy problems and is meant to provide an idealized overview for the different requirements related to different combinations of complexity, uncertainty, and ambiguity. Under real conditions, energy problems and their conditions often turn out to be more interdependent among each other and the required measures more contingent on the context in which the policy is embedded.

3.6.2 Formats for the three different discourse types

3.6.2.1 Instruments for the instrumental and epistemic discourse (pool 1)

Resolving conflicts in situations characterized by linear problems requires deliberation among knowledge carriers. The instruments listed in this category provide opportunities

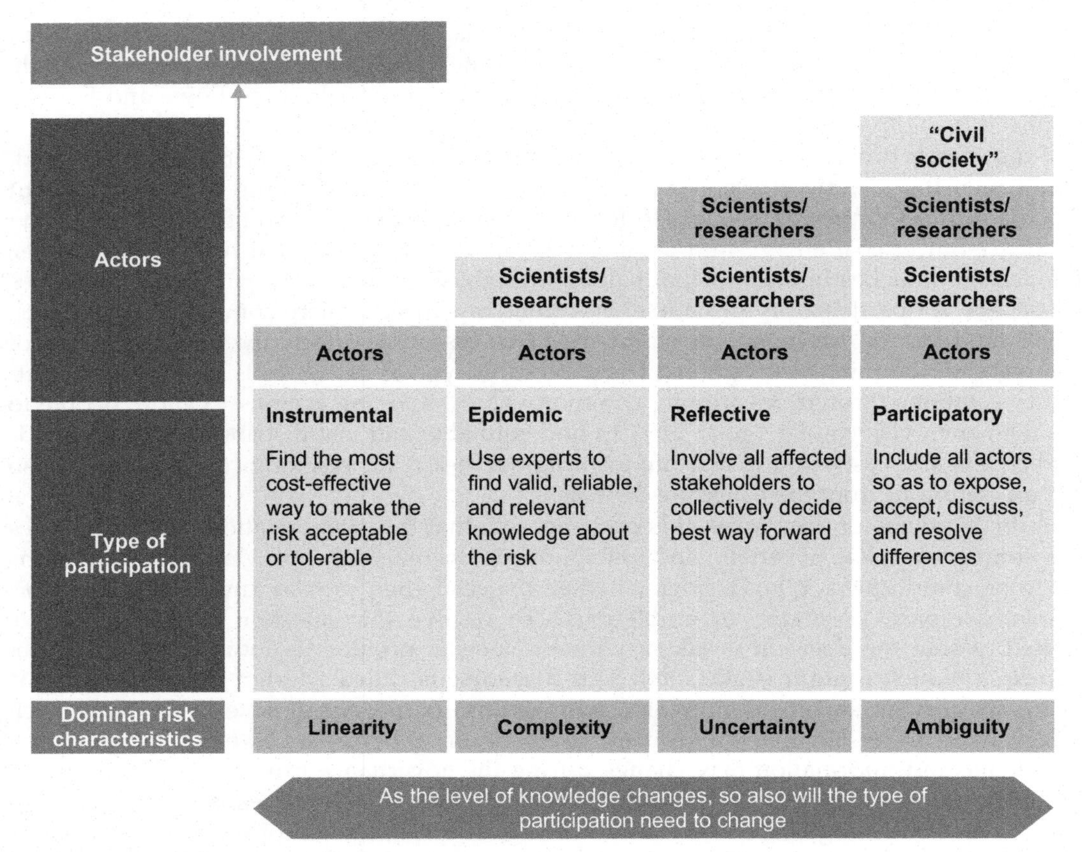

FIGURE 3.2 The risk-management escalator and stakeholder involvement. Source: *Reproduced from IRGC (International Risk Governance Council), 2005. Risk Governance – Towards an Integrative Approach, White Paper No 1, IRGC, Geneva, p. 53.*

for experts (not necessarily scientists) to argue over the factual impact assessment with respect to the energy issue in question. The objective of a discourse in each of these instruments is the most adequate description or explanation of a phenomenon (e.g., the question: which impacts are to be expected by implementing a specific policy?). Instruments that promise to meet these requirements are expert hearings, expert workshops, science workshops, expert panels or advisory committees, or consensus conferences as practiced in the medical field (Roqueplo, 1995; Koenig and Jasanoff, 2001). If anecdotal knowledge is needed, one can refer to focus groups, panels of volunteers, and simple surveys (Dürrenberger et al., 1999). More sophisticated methods of reducing complexity for difficult risk issues include Delphi methods, group Delphi, metaanalytical workshops, and scoping exercises (Webler et al., 1991; Sutton et al., 2000). The most frequently used instruments for resolving epistemic conflicts are described next (OECD (Organisation for Economic Co-operation and Development), 2002):

Expert hearing: This is the most popular form of resolving differences among experts (Boehmer-Christiansen, 1997; OECD (Organisation for Economic Co-operation and Development), 2002). Experts with different positions are asked to testify before the representatives of the organizing institution (most often a regulatory agency) or the deliberative panel. The organizers ask each expert a specific question and let them develop their line of arguments. Occasionally, hearings allow for open discussions among the experts, but the final judgment is left to the organizing committee or the deliberative panel. Hearings are excellent and fairly inexpensive settings, if the objective is to get a clearer picture of the variability of expert judgments and to become aware of the arguments supporting each position. Hearings do not provide consensus and may not resolve any conflict. However, they may clarify the basis of the conflict or the different points of view in a contested risk issue.

Expert committees: Expert committees, advisory boards, think tanks, and scientific commissions are also very popular forms of involving external knowledge carriers within the risk-management process (Rakel, 2004). They have the advantage that experts interact freely with each other, have more time to learn from each other, and are able to consult other experts if deemed necessary. They work independently of the agency or deliberative body to which they report. The main disadvantage is that expert committees may not arrive at a consensus, may take too much time to reach a conclusion, may not respond to the urgent needs of the deliberative body, and may "live a life of their own." In addition, many expert committees can only come to an agreement if the members have similar backgrounds and positions.

Expert consensus conference: Particularly in the medical field, experts are gathered in a workshop to discuss treatment options and to decide on a general standard to be applied in comparable cases throughout the world (Jones and Hunter, 1995). The workshop is organized in group sessions in order to prepare common standards and in plenary sessions to reach a common agreement. One could envision consensus conferences in the risk area for the purpose of setting and articulating common conventions for risk assessment and evaluation.

Delphi exercises: A Delphi process is aimed at obtaining a wide range of opinions among a group of experts (Linstone and Turoff, 2002). The process is organized in four steps: in step 1 a questionnaire asks a group of distinguished scientists to assess the severity or the scope of a risk. The scientists provide their best assignments, possibly including some type of uncertainty interval to their answers. In step 2 the organizing team feeds back to each participant the scores of the whole group, including medians, standard deviation, and aggregated uncertainty intervals. Each individual is then asked to perform the same task again, but now with the knowledge of the responses of all other participants. In step 3, this procedure is repeated until individuals do not change their assessment any more. In step 4 the organizer summarizes the results and articulates the conclusions. A variation of the classic Delphi method is the Group Delphi (Webler et al., 1991). During a Group Delphi, all participants meet face to face and make the assessments in randomly assigned small groups of three and four. The groups whose average scores deviate most from the median of all other groups are requested to defend their position in a plenary session. Then the small groups are reshuffled and perform the same task again. This process can be iterated three or four times until no further significant changes are made. The advantage of Delphi is that a serious effort has been invested in finding the common ground among the experts and in finding the reasons and arguments that cause differences in assessments. The disadvantage is that the quality of Delphi outcomes depends upon the accuracy and completeness of the expertise and information brought into the process.

3.6.2.2 Instruments for the reflective discourse (pool 2)

The next set of instruments refers to risk with high uncertainty focusing on the impacts when estimating the likely impacts of various energy policy options. Scientific input is also needed in order to compile the relevant data and the various arguments for the positions of the different science camps. Procedures such as the "pedigree scheme" by Funtowicz and Ravetz (1990) might be helpful in organizing the existing knowledge. Furthermore, information about the different types of uncertainties has to be collected and brought into a deliberative arena. The central objective is, however, to deliberate about the postprudent handling of unresolved uncertainty. For this purpose, representatives of affected stakeholders and public interest groups must be identified, invited, and informed about the issue in question (Yosie and Herbst, 1998: 644ff). The objective of the deliberation is to find the right balance between too little and too much precaution. There is no scientific answer to this question, and even economic balancing procedures are of limited value since the stakes are uncertain. Major instruments for reflective discourses are roundtables, negotiated rulemaking, mediation, arbitration, and stakeholder dialogues. The most popular instruments for conducting a reflective discourse within a deliberative setting are the following (OECD (Organisation for Economic Co-operation and Development), 2002):

Stakeholder hearings: Most regulatory regimes of the world require hearings with stakeholders or directly affected citizens under specific circumstances (Renn et al., 1995). Such hearings can serve a useful purpose if they are meant to give external stakeholders the opportunity to voice their opinion and arguments. Hearings also provide opportunities for stakeholders to understand the position of the regulatory agencies or other direct players (such as industry). But hearings have proven very ineffective for resolving conflicts or pacifying heated debates. On the contrary, hearings normally aggravate the tone of the conflict and lead to polarizations. Other than for the purpose of investigating, the concerns and objections of organized groups, stakeholder hearings should be avoided.

Roundtables (advisory committees, stakeholder dialogues, and negotiated rulemaking): Roundtables are very popular settings for stakeholder involvement (US EPA/SAB, 2001; Stoll-Kleemann and Welp, 2006). Normally, the participants represent the major social groups, such as employers, unions, and professional associations. The advantage is that the ritual window-dressing activities (typical for classic hearings) can be overcome through the continuity of the process and a strict working atmosphere. The major disadvantage is that groups outside the roundtable and representatives of the general public are left out. They can only trust the process to be effective and fair. If the debate is heated and adversarial elements govern the political climate, roundtables will face severe difficulties to legitimize their agreements. For many regulatory issues and risk-management decisions, however, such roundtables have been very effective and also cost-efficient in incorporating the perspective of organized groups and in suggesting adequate management options. There are also good techniques available (such as value-tree analysis, multiattribute decision-structuring, and metaplanning exercises) to make these heterogeneous group meetings more productive (Rauschmayer and Wittmer, 2006). Essential for organizing a successful roundtable is the involvement of a professional moderator. Moderation should be performed by a neutral institution rather than the organizer.

Mediation (arbitration and alternate dispute resolution methods): If conflicts are already clearly visible and unavoidable, the procedures of alternate dispute resolution are effective and less costly instruments compared to legal litigation (Hadden, 1995; US EPA, 1995; Susskind and Fields, 1996). Mediation and similar procedures rest on the assumption that stakeholders can find a common solution if they do not insist on positions but try to meet their crucial interests and underlying values. Under these circumstances, win–win solutions may be available that will please all parties. Mediation requires the involvement of a skilled and professional mediator. Similar to roundtables, such mediators should be recruited from neutral professional services. It is advisable that mediators have sufficient knowledge about the issue, that they can understand and evaluate all participants' statements, but that they do not have a clear commitment to one or the other side. The advantage of mediation is that conflicts among participants can be reconciled before they reach the legal arena. The disadvantage is that, depending upon the composition of the group, interests which are not emphasized at the roundtable will not be considered. Most alternate dispute resolution methods work well under the condition of adversarial and corporatist styles; they may be seen as unnecessary in more trustful environments where conflicts are rare and stakeholders less agitated.

3.6.2.3 Instruments for the participatory discourse (pool 3)

The last group of instruments addressed in this subsection deals with ambiguity. Most often, ambiguities arise over the issue of social or moral justification of a policy including the decision of who is to be responsible and accountable. Before investing in resolving ambiguity, it is essential to investigate the cause of the ambiguity and to find the right spot where the involvement procedure would best fit in the decision-making process. Preferred instruments here are citizen panels or juries (randomly selected), citizen advisory committees or councils, public consensus conferences, citizen action groups, and other participatory techniques. The main instruments belonging to this category are as follows (OECD (Organisation for Economic Co-operation and Development), 2002):

Public hearings: Public hearings are required in many regulatory regimes all over the world (Renn et al., 1995). The idea is that people who feel affected by a decision should be given an opportunity to make their concerns known to the authorities and, vice versa, to give the authorities the opportunity to explain their viewpoint to the public. Although public hearings are fairly inexpensive and easy to organize; their effectiveness is rated as poor in most of the scientific investigations on the subject. Hearings tend to stereotype the issue and the actors involved, to aggravate emotions, to emphasize dissent rather than consensus, and to amplify distrust rather than generate trust. Unless the issue is only slightly controversial and the climate is characterized by an overall consensual mood, we do not recommend public hearings as a setting for resolving ambiguity.

Surveys and focus group: Surveys of the general public or special groups are excellent settings in which to explore the concerns and worries of the addressed audience (Dürrenberger et al., 1999). If they are performed professionally, the results are usually valid and reliable. The results of surveys provide, however, only a temporary snap shot of public opinion; they do not produce solutions for conflict resolution or predict the fate of positions once they have entered the public arena. Surveys describe the starting position before a conflict may unfold. Focus groups go one step further by exposing arguments to counterarguments in a small group discussion setting. The moderator introduces a stimulus (e.g., statements about the risk) and lets members of the group react to the stimulus and to each other's statements. Focus groups provide more than data about people's positions and concerns; they also measure the strength and social resonance of each argument vis-à-vis counterarguments. The major disadvantage of surveys and focus groups is the lack of real interaction among participants. Therefore both instruments are advisable as preliminary steps in understanding the context and the expectations; but they do not assist risk managers in resolving a pressing issue. In addition, both instruments are fairly expensive.

Citizen advisory committees (ombudsman, neighborhood associations, and citizen boards): The instrument of citizen advisory committees is particularly popular in local and regional contexts, but there are also examples for advisory committees on a national level (Vari, 1995; Applegate, 1998). The chemical industry has been experimenting with citizen advisory committees for a long time in the framework of its responsible care program (Prakash, 2000). This program is directed toward people in the vicinity of chemical installations. Such an approach is also feasible with consumers if companies or agencies would like to involve their ultimate clients in the risk-management process. The problem here is selection: either one invites representatives of stakeholder groups (such as the consumer associations) or one tries to find a sample of "representative" consumers of the specific products or chemicals under review. Both approaches have their merits and drawbacks. Stakeholder groups are often quite distanced from the members they are supposed to represent. This is particularly true for consumer associations since consumers form a very heterogeneous group, and the majority of them do not belong to consumers associations. At the same time a representative sample of consumers is difficult to obtain and it is questionable whether such a sample can speak in the interest of all consumers. In spite of these difficulties, such advisory committees can be very effective in detecting potential conflicts (early warning function) and getting the concerns of the consumers heard and reflected in the respective organizing institutions. In addition, the organization of citizen advisory committees is fairly inexpensive and easy to do.

Citizen consensus conferences: The Danish Board of Technology introduced a new form of citizen involvement, which it called "consensus conferencing." This instrument is strongly based on the belief that a group of noncommitted and independent citizens is best to judge the acceptability or tolerability of technological risks (Joss, 1998; Sclove, 1995; Andersen and Jaeger, 1999). Six to ten citizens are invited to study a risk issue in detail and to provide the legal decision-maker, or an agency, with a recommendation at the end of the process. The citizens are usually recruited by self-selection. The organizers put an advertisement in the newspaper asking for volunteers. If too many people volunteer for the consensus conference, the organizers follow specific rules for ensuring equal representation. An equal number of women and men are required, as well as a cross section of the population in terms of age, social class, and political preferences. The participants receive a substantial amount of material before they convene for the first time. They study this material during two consecutive weekends. The consensus conference itself lasts 3 days. During the first day the participants share their reflections with a body of regulators or decision-makers (often members of parliament). They also raise their questions and listen to the answers given by politicians and experts. On the second day in the morning the hearing continues, but this time it is open to the wider public. In the afternoon the participants meet behind closed doors and articulate their recommendations. These are then presented to the decision-makers on the following day. The decision-makers have the opportunity to give further comments. Finally, the participants write the final draft of the recommendations and present them to the media at the end of the third day. The advantage of consensus conferencing is the transposition of a major conflict to a small group of laypeople who are being educated about the subject and are asked to make recommendations based on their knowledge and personal values. The main disadvantage is the small number of people who are assigned such an important task. The restricted number of 6–10 participants has been the thrust of criticism in the literature (Einsiedel and Eastlick, 2000). Consensus conferences seem to yield a compelling legitimacy effect within countries that are small and emphasize consensus over conflict. Most successful trials are reported in Denmark, Norway, and Switzerland. The experiences in more adversarial countries such as the United Kingdom, France, and Germany are less encouraging (Joss, 1997). The results of the deliberations were not widely published in the media; decision-makers were not willing to submit sufficient time to small groups of laypeople; and administrators paid only lip service to the conference statements.

Citizen panels, planning cells or citizen juries: Planning cells or citizen panels (juries) are groups of randomly selected citizens who are asked to compose a set of policy recommendations on a specific issue (Crosby et al., 1986; Dienel, 1989; Stewart et al., 1994). The objective is to provide citizens with the opportunity for learning about the technical and political facets of the risk-management options and for enabling them to discuss and evaluate these options and their likely consequences according to their own set of values and preferences. The participants are informed about the potential options and their consequences before they are asked to evaluate these options. Since the process requires time for the educational program and the evaluation of options, the panels are conducted in seminar form over 3–5 consecutive days. All participants are exposed to a standardized program of information, including hearings, lectures, panel discussions, videotapes, and field tours. Since participants are selected by random procedures, every individual in the affected population has an equal opportunity to participate in the process. In reality, however, only 5%–40% of the randomly selected citizens decide to become active participants. In contrast to consensus conferences, however, the number of people who can participate is limited only by available resources and time. Several hundred citizens can be involved in one exercise. All participants are grouped in panels of 20–25, with an identical educational program and evaluative tasks. If most of the panels come up with similar conclusions, one can be sure that this is (or would be) the will of the informed public. Planning cells require a large investment of time and money and are not suitable for all types of problems and all contexts. If the problem is highly technical, it may be impossible to bring citizens up to the necessary level of understanding. Furthermore, if the decision options are too narrowly restricted and there is not enough room to allow trade-offs on decision criteria, the process will fail. Citizen panels may also face the problem of being legitimate consultants to policy makers in an adversarial climate.

3.6.2.4 *Synthesis of components: design discourse*

The three pools of instruments provide a sufficient number of choices for matching the instrument with the energy issue at hand. It is more difficult, however, to find the right and appropriate combination if instruments from several pools must be combined. For example, if the policy options are characterized by high complexity, uncertainty, and ambiguity, one needs an integrated process that includes a sequential chain consisting of at least one instrument from each pool. Or if the issue is highly complex and uncertain but raises little controversy, only instruments from pools 1 or 2 have to be selected. The pools provide tool boxes and, depending upon the nature of the energy issue and what policy makers know about them, the selection has to be made from one or more of the three pools.

The selection of a specific sequence and the choice of instruments depend upon the nature of the issue, but also upon the context and the regulatory structure and culture of the country or state in which such a process is planned. Different countries have developed diverse traditions and different preferences when it comes to deliberative processes (Löfstedt and Vogel, 2001). The selection of instruments from the three pools, therefore, needs to reflect the nature of the problem, the regulatory system and the respective political culture.

One possibility for a policy making body, for example, a national energy agency, for selecting one instrument from each pool might be to organize an expert hearing to understand the knowledge claims that are relevant for the issue. Afterwards, the agency might initiate a roundtable for negotiated rulemaking in order to find the appropriate trade-offs between the costs of overregulation and underregulation in the face of major uncertainties. The agency could then convene several citizen advisory committees to explore the potential conflicts and dissenting interpretations of the situation and the proposed policy options. Each component builds upon the results of the previous component.

However, the requirements of an integrated participation program are not met by merely running three or more different components in parallel and feeding the output of each component as input into the next component (Renn, 2008). Each type of discourse has to be embedded in an integrated procedural concept. In addition, there is a need for continuity and consistency throughout the whole process. Several participants may be asked to take part in all three instruments, and an oversight committee may be necessary to provide the mandated links. One example for such an integrated hybrid model is the "cooperative discourse" approach (Renn, 1999). The three pools of discourse organizations are summarized in Table 3.2.

3.6.3 The model of cooperative discourse

The model of "cooperative discourse" is one hybrid model that integrates the three poops and is based on a combination of functional and deliberative concept of participation (Renn, 2008: 343ff). The model is based on three steps:

- knowledge based on technical expertise (epistemic discourse);
- knowledge and values derived from social interests and advocacy (reflective discourse);
- knowledge and values based on common sense, folklore wisdom, and personal experience (participatory discourse).

TABLE 3.2 Appropriate formats for different risk discourse types.

	Challenge	Objective	Function	Instruments
Type 1—epistemic	Complexity	Inclusion of best available knowledge	Agreement on casual relations and effective measures	Expert panels, expert hearings, metaanalysis, Delphi, etc.
Type 2—reflective	Uncertainty	Fair and acceptable arrangement for benefit and burden sharing	Balancing costs of underprotection versus costs of overprotection facing uncertain outcomes	Negotiated rulemaking, mediation, roundtables, stakeholder meetings, etc.
Type 3—participatory	Ambiguity	Congruency with social and cultural values	Resolving value conflicts and assuring fair treatment of concerns and visions	Citizen advisory committees, citizen panels, citizen jury, consensus conferences, public meetings, etc.
Hybrid designs	Combination	Meeting more than one challenge	Meaningful and effective integration of functions	Selection from each of the three types

These three forms of knowledge and values are integrated into a sequential procedure of three consecutive steps. The *first step* refers to the identification of objectives or goals that the process should reflect. The identification of concerns and objectives is best accomplished by asking all relevant stakeholder groups (i.e., socially organized groups that are, or perceive themselves as being, affected by the decision) to reveal their values and criteria for judging different options (reflective discourse). This can be done by a process called value-tree analysis which provides a systematic overview of stakeholder concerns and values (Baudry et al., 2018).

With different policy options and criteria available, the *second step* involves the actors with special knowledge and evidence on the subject. Experts representing multiple disciplines and plural viewpoints about the issue in question are asked to judge the performance of each option on each indicator (epistemic discourse). For this purpose, the group Delphi method is used (Webler et al., 1995; Wassermann et al., 2011).

The *third and last step* is the evaluation of each option profile by one group, or several groups of randomly selected citizens, in a participatory discourse. We refer to these panels as citizen panels for policy evaluation and recommendation. The objective is to provide citizens with the opportunity for learning about the technical and political facets of policy options and for enabling them to discuss and evaluate these options and their likely consequences according to their own set of values and preferences. The idea is to conduct a process loosely, analogous to a jury trial with experts and stakeholders as witnesses and advisers on procedure as "professional" judges (Renn, 2015).

The whole process is supervised by a group of official policy makers, major stakeholders, and moral agents (such as religious or cultural leaders). Their task is to oversee the process, test and examine the material handed out to the panelists, review the decision frames and questions, and write a final interpretation of the results.

The applications of cooperative discourse method provide some evidence and reconfirmation that the theoretical expectations linked to the combination of epistemic, reflective, and participatory discourses can be met on the local, regional, and national level.

Evaluation studies by independent scholars confirmed that the objectives of effectiveness, efficiency, and social acceptability were largely met in the cases in which this hybrid model was used (Roch, 1997; Vorwerk and Kämper, 1997).

3.7 Conclusion

The objective of this chapter has been to address and discuss the need and potential for stakeholder involvement in energy policy making and to introduce different conceptual models of how to respond to the two crucial questions of any involvement process: first, who and what should be included? Second, what kind of output should the involvement process produce in order to facilitate better decision-making in complex issues such as energy policy making? The chapter explained the requirements for involvement processes when approaching complex, uncertain, and ambiguous energy problems. Such problems require epistemic, reflective, and participatory discourse structures. For each of these discourse types, appropriate formats are available that can be used to address the specific energy issue under investigation.

There has been much concern in the literature that opening the energy policy arena to stakeholder input would lead to a dismissal of factual knowledge and to inefficient spending of public money (Sunstein, 2002). Given the experience with stakeholder involvement so far, these concerns are not warranted. There are only a few voices that wish to restrict scientific input to energy policy making. Scientific expertise is an essential element of stakeholder involvement and a crucial pillar of all formats for stakeholder involvement. The role of scientific analysis in designing and evaluating energy policies should not be weakened but, rather, strengthened when opening the arena for stakeholder input.

Profound scientific knowledge is required especially with regard to dealing with complexity. This knowledge has to be assessed and collected by scientists and energy professionals who are recognized as competent authorities in the respective field. The systematic search for the "state-of-the-art" in scientific analysis and oversight leads to a knowledge base that provides the data for deliberation. At the same time, however, the style of deliberation also should transform the scientific discourse and lead the discussion toward classifying knowledge claims, characterizing uncertainties, exploring the range of alternative explanations, and acknowledging the limits of systematic knowledge in many policy arenas. This can be done in any country, independent of political system, or governmental structure. Stakeholder involvement and public participation have been used and successfully implemented in many developing countries and threshold countries such as China (Grimble and Chan, 1995; Tang et al., 2005).

Once the potential contributions of the expert communities, the stakeholder groups, and members of the affected public had been recognized and acknowledged in an adequate combination of formats, a process of mutual understanding and constructive decision-making started to unfold (Pahl-Wostl, 2002). Such a discursive process may not always lead to the desired results, but the experiences so far justify a fairly optimistic outlook. The main lesson from these experiences has been that scientific expertise, rational decision-making, and public values can be reconciled if a serious attempt is made to integrate them. The transformation of the energy policy arena into a well-structured and professionally moderated analytic–deliberative discourse seems to be an essential and, ultimately, inevitable step toward improving energy policies and facilitating the transformation toward a sustainable energy future.

References

Abels, G., 2007. Citizen involvement in public policy–making: does it improve democratic legitimacy and accountability? The case of pTA. Interdiscip. Inf. Sci. 13 (1), 103–116.

Ackerman, B., Fishkin, J.S., 2004. Deliberation Day. Yale University Press, New Haven, CT.

Alexander, J., 1993. The return of civil society. Contemp. Soc. 22, 797–803.

Amy, D.J., 1983. Environmental mediation: an alternative approach to policy stalemates. Policy Sci. 15, 345–365.

Amy, D.J., 1987. The Politics of Environmental Mediation. Cambridge University Press, Cambridge and New York.

Andersen, I.-E., Jaeger, B., 1999. Scenario workshops and consensus conferences. Towards more democratic decision-making. Sci. Public Policy 26, 331–340.

Applegate, J., 1998. Beyond the usual suspects: the use of citizens advisory boards in environmental decision-making. Indiana Law J. 73, 903.

Baudry, G., Macharis, C., Valee, T., 2018. Range-based multi-actor multi-criteria analysis: a combined method of multi-actor multi-criteria analysis and Monte Carlo simulation to support participatory decision-making under uncertainty. Eur. J. Oper. Res. 264 (1), 257–269.

Baughman, M., 1995. Mediation. In: Renn, O., Webler, T., Wiedemann, P. (Eds.), Fairness and Competence in Citizen Participation: Evaluating New Models for Environmental Discourse. Kluwer, Dordrecht and Boston, pp. 253–266.

Benhabib, S., 1992. Autonomy, modernity, and community: communitarianism and critical theory in dialogue. In: Honneth, A., McCarthy, T., Offe, C., Wellmer, A. (Eds.), Cultural-Political Interventions in the Unfinished Project of Enlightenment. MIT Press, Cambridge, pp. 39–61.

Boehmer-Christiansen, S., 1997. Reflections on scientific advice and EC transboundary pollution policy. Sci. Public Policy 22 (3), 195–203.

Brauers, H., Herpich, P., von Hirschhausen, C., Jürgens, I., Neuhoff, K., Oei, P.-Y., et al., 2018. Coal Transition in Germany – Learning From Past Transitions to Build Phase-Out Pathways. IDDRI and Climate Strategies, Paris and Berlin.

Chambers, S., 2003. Deliberative democratic theory. Annu. Rev. Pol. Sci. 6, 307–326.

Coglianese, C., 1997. Assessing consensus: the promise and performance of negotiated rule-making. Duke Law J. 46, 1255–1333.

Crosby, N., Kelly, J.M., Schaefer, P., 1986. Citizen panels: a new approach to citizen participation. Public Adm. Rev. 46, 170–178.

De Marchi, B., Ravetz, J.R., 1999. Risk management and governance: a post-normal science approach. Futures 31, 743–757.

Dewey, J., 1940. The Public and Its Problems: An Essay in Political Inquiry. Gateway, Chicago, IL.

Dienel, P.C., 1989. Contributing to social decision methodology: citizen reports on technological projects. In: Vlek, C., Cvetkovich, G. (Eds.), Social Decision Methodology for Technological Projects. Kluwer Academic, Dordrecht, pp. 133–151.

Dryzek, J.S., 1994. Discursive Democracy: Politics, Policy, and Political Science, second ed. Cambridge University Press, Cambridge.

Dürrenberger, G., Kastenholz, H., Behringer, J., 1999. Integrated assessment focus groups: bridging the gap between science and policy? Sci. Public Policy 26 (5), 341–349.

Einsiedel, E.F., Eastlick, D.L., 2000. Consensus conferences as deliberative democracy: a communications perspective. Sci. Commun. 21, 323–343.

Ethridge, M.E., 1987. Procedures for citizen involvement in environmental policy: an assessment of policy effects. In: DeSario, J., Langton, S. (Eds.), Citizen Participation in Public Decision Making. Greenwood Press, Westport, pp. 115–132.

Fischer, F., 2005. Participative governance as deliberative empowerment. The cultural politics of discursive space. Am. Rev. Public Adm. 36 (1), 19–40.

Fisher, R., Ury, W., 1981. Getting to Yes: Negotiating Agreement Without Giving In. Penguin Books, New York.

Fiorino, D.J., 1990. Citizen participation and environmental risk: a survey of institutional mechanisms. Sci. Technol. Hum. Values 15 (2), 226–243.

Foucault, M., 2003. Die Ordnung der Dinge, special ed. Suhrkamp, Frankfurt am Main.

Fung, A., Wright, E.O., 2001. Deepening democracy: innovations in empowered local governance. Pol. Soc. 29 (1), 5–41.

Funtowicz, S.O., Ravetz, J.R., 1990. Uncertainty and Quality in Science for Policy. Kluwer, Dordrecht.

Granovetter, M., 1992. Economic action and social structure: the problem of embeddedness. In: Granovetter, M., Swedberg, R. (Eds.), The Sociology of Economic Life. Westview Press, Boulder and Oxford, pp. 53–81.

Gregory, R., McDaniels, T., Fields, D., 2001. Decision aiding, not dispute resolution: a new perspective for environmental negotiation. J. Policy Anal. Manage. 20 (3), 415–432.

Grimble, R., Chan, M.K., 1995. Stakeholder analysis for natural resources management in developing countries. Some practical guidelines for making management more participatory and effective. Nat. Resour. Forum 19 (2), 113–124.

Habermas, J., 1968. Technik und Wissenschaft als Ideologie. Suhrkamp, Frankfurt.

Habermas, J., 1970. Towards a theory of communicative competence. Inquiry 13, 363–372.

Habermas, J., 1984. Theory of Communicative Action. Volume 1: Reason and the Rationalization of Society. Beacon Press, Boston, MA.

Habermas, J., 1992. Faktizität und Geltung – Beiträge zur Diskurstheorie des Rechts und des demokratischen Rechtsstaats, second ed. Suhrkamp, Frankfurt am Main.

Habermas, J., 1996. Die Einbeziehung des Anderen – Studien zur politischen Theorie. Suhrkamp, Frankfurt am Main.

Hadden, S., 1995. Regulatory negotiation as citizen participation: a critique. In: Renn, O., Webler, T., Wiedemann, P. (Eds.), Fairness and Competence in Citizen Participation: Evaluating New Models for Environmental Discourse. Kluwer, Dordrecht and Boston, pp. 239–252.

Hagendijk, R., Irwin, A., 2006. Public deliberation and governance: engaging with science and technology in contemporary Europe. Minerva 44, 167–184.

Hammer, G., 2003. American Pragmatism: A Religious Genealogy. Oxford University Press, Oxford.

Heuberger, C.F., Mac Dowell, N., 2018. Real-world challenges with rapid transition to 100% renewable power systems. Joule 2 (3), 367–370.

Hillmann, K.-H., 1994. Wörterbuch der Soziologie, fourth ed. Kröner, Stuttgart.

Horlick-Jones, T., Rowe, G., Walls, J., 2007. Citizen engagement processes as information systems: the role of knowledge and the concept of translation quality. Public Underst. Sci. 16, 259–278.

Hostmann, M., Bernauer, T., Mosler, H.-J., Reichert, P., Truffer, B., 2005. Multi-attribute value theory as a framework for conflict resolution in river rehabilitation. Journal of Multi-Criteria Decision Analysis 13, 91–102.

IRGC (International Risk Governance Council), 2005. Risk Governance – Towards an Integrative Approach. IRGC, Geneva, White Paper No 1.

IRGC (International Risk Governance Council), 2017. Risk Governance – Revised Version. IRGC, Geneva.

Jaeger, C., Renn, O., Rosa, E., Webler, T., 2001. Risk, Uncertainty and Rational Action. Earthscan, London.

Jasanoff, S., 1993. Bridging the two cultures of risk analysis. Risk Anal. 13 (2), 123–129.

Jasonoff, S., 1998. The political science of risk perception. Reliab. Eng. Syst. Saf. 59, 91–99.

Jones, J., Hunter, D., 1995. Qualitative research: consensus methods for medical and health services research. Br. Med. J. 311 (7001), 376–380.

Joss, S., 1997. Experiences with consensus conferences. In: Paper Presented to the International Conference on Technology and Democracy, Centre for Technology and Culture, University of Oslo, Norway, 17–19 January 1997, Science Museum, London.

Joss, S., 1998. Danish consensus as a model of participatory technology assessment: an impact study of consensus conferences on Danish Parliament and Danish Public Debate. Sci. Public Policy 25, 2–22.

Joss, S., 2005. Lost in translation? Challenges for participatory governance of science and technology. In: Bogner, A., Torgersen, H. (Eds.), Wozu Experten? Ambivalenzen der Beziehung von Wissenschaft und Politik. VS Verlag für Sozialwissenschaften, Wiesbaden, pp. 197–219.

Klinke, A., Renn, O., 2012. Adaptive and integrative governance on risk and uncertainty. J. Risk Res. 15 (3), 273–292.

Klinke, A., Renn, O., 2019. The coming age of risk governance. Risk Anal. published online July 11, 2019. Available from: https://doi.org/10.1111/risa.13383.

Koenig, A., Jasanoff, S., 2001. A Comparison of Institutional Changes to Improve the Credibility of Expert Advice for Regulatory Decision Making in the US and EU. Study for the European Commission, EU, Brussels.

Larsen, L., Harlan, S.L., Bolin, B., Hacket, E.J., Hope, D., Kirby, A., et al., 2004. Bonding and bridging: understanding the relationship between social capital and civic action. J. Plan. Educ. Res. 24 (1), 64–77.

Lauber, V., Jacobsson, S., 2015. Lessons from Germany's Energiewende. In: Fagerberg, J., Laesradius, S., Martin, B.R. (Eds.), The Triple Challenge for Europe: Economic Development, Climate Change, and Governance. University Press, Oxford, pp. 173–203.

Linnerooth-Bayer, J., Fitzgerald, K.B., 1996. Conflicting views on fair siting processes evidence from Austria and the US. Risk: Health Saf. Environ. 7 (2), 119–134.

Linstone, H.A., Turoff, M. (Eds.), 2002. The Delphi Method. Techniques and Applications. ISA Applications, Newark. Available from: www.is.njit.edu/pubs/delphibook/.

Löfstedt, R., Vogel, D., 2001. The changing character of regulation: a comparison of Europe and the United States. Risk Anal. 21 (3), 393–402.

Luhmann, N., 1983. Legitimation durch Verfahren. Suhrkamp, Frankfurt am Main.

Luhmann, N., 1989. Ecological Communication. Polity Press, Cambridge.

Malinowski, B.A., 1944. Scientific Theory of Culture. University of North Carolina Press, Chapel Hill.

McCormick, S., 2007. Democratizing science movements: a new framework for contestation. Soc. Stud. Sci. 37, 1–15.

McDaniels, T., 1996. The structured value referendum: eliciting preferences for environmental policy alternatives. J. Policy Anal. Manage. 15 (2), 227–251.

Merton, R.K., 1959. Social Theory and Social Structure: Toward the Codification of Theory and Research. The Free Press, Glencoe, IL (originally published 1949).

Münch, R., 1982. Basale Soziologie: Soziologie der Politik. Suhrkamp, Frankfurt am Main.

Nanz, P., Leggewie, C., 2019. No Representation Without Consultation. A Citizens Guide to Participatory Democracy, Between-the-lines Publ, Toronto.

OECD (Organisation for Economic Co-operation and Development), 2002. Guidance Document on Risk Communication for Chemical Risk Management. OECD, Paris.

Pahl-Wostl, C., 2002. Participative and stakeholder-based policy design, analysis and evaluation processes. Integr. Assess. 3 (1), 3–14.

Parsons, T.E., 1951. The Social System. Free Press, Glencoe.

Parsons, T., 1967. Sociological Theory and Modern Society. Free Press, New York.

Parsons, T., 1971. The System of Modern Societies. Prentice-Hall, Englewood Cliffs, NJ.

Parsons, T.E., Shils, E.A., 1951. Toward a General Theory of Action. Cambridge University Press, Cambridge.

Pierce, C.S., 1867. On the natural classification of arguments. Proc. Am. Acad. Arts Sci. 7, 261–287.

Prakash, A., 2000. Responsible care: an assessment. Bus. Soc. 39 (2), 183–209.

Radcliffe-Brown, R., 1935. On the concept of function in social science. Am. Anthropologist 37, 394–402.

Raiffa, H., 1994. The Art and Science of Negotiation, 12th ed. Cambridge University Press, Cambridge.

Rakel, H., 2004. Scientists as expert advisors: science cultures versus national cultures? In: Kurz-Milcke, E., Gigerenzer, G. (Eds.), Experts in Science and Society. Springer, Heidelberg, Berlin, New York, pp. 3–25.

Rauschmayer, F., Wittmer, H., 2006. Evaluating deliberative and analytical methods for the resolution of environmental conflicts. Land. Use Policy 23, 108–122.

Renn, O., 1999. A model for an analytic–deliberative process in risk management. Environ. Sci. Technol. 33 (18), 3049–3055.

Renn, O., 2004. The challenge of integrating deliberation and expertise: participation and discourse in risk management. In: MacDaniels, T.L., Small, M.J. (Eds.), Risk Analysis and Society: An Interdisciplinary Characterization of the Field. Cambridge University Press, Cambridge, pp. 289–366.

Renn, O., 2008. Risk Governance. Coping With Uncertainty in a Complex World. Earthscan, London.

Renn, O., 2010. The contribution of different types of knowledge towards understanding, sharing and communicating risk concepts. Catalan J. Commun. Cult. Stud. 2 (2), 177–195.

Renn, O., 2014. Stakeholder Involvement in Risk Governance. Ark Publications, London.
Renn, O., 2015. Are we afraid of the wrong things? Statistics, psychology and the risk paradox, in. In: Union Investment (Ed.), The Measurement of Risk. Union Investment International, Frankfurt, pp. 24–35.
Renn, O., Marshall, J.P., 2016. Coal, nuclear and renewable energy policies in Germany: From the 1950 to the "Energiewend. Energy Policy 99 (C), 224–232.
Renn, O., Schweizer, P., 2009. Inclusive risk governance: concepts and application to environmental policy making. Environ. Policy Gov. 19 (3), 174–185.
Renn, O., Webler, T., Rakel, H., Dienel, P.C., Johnson, B., 1993. Public participation in decision making: a three-step procedure. Policy Sci. 26, 189–214.
Renn, O., Webler, T., Wiedemann, P., 1995. The pursuit of fair and competent citizen participation. In: Renn, O., Webler, T., Wiedemann, P. (Eds.), Fairness and Competence in Citizen Participation: Evaluating New Models for Environmental Discourse. Kluwer, Dordrecht and Boston, pp. 339–368.
Renn, O., Klinke, A., van Asselt, M., 2011. Coping with complexity, uncertainty and ambiguity in risk governance: a synthesis. AMBIO 40 (2), 231–246.
Roch, I., 1997. Evaluation der 3. Phase des Bürgerbeteiligungsverfahrens in der Region Nordschwarzwald. Akademie für Technikfolgenabschätzung, Stuttgart, Research Report No 71.
Roqueplo, P., 1995. Scientific expertise among political powers, administrators and public opinion. Sci. Public Policy 22 (3), 175–182.
Rosa, E.A., Renn, O., McCright, A.M., 2014. The Risk Society Revisited. Social Theory and Governance. Temple University Press, Philadelphia, PA.
Rowe, G., Frewer, L.J., 2000. Public participation methods: a framework for evaluation. Sci., Technol. Hum. Values 225 (1), 3–29.
Rowe, G., Marsh, R., Frewer, L.J., 2004. Evaluation of a deliberative conference. Sci., Technol., Hum. Values 29 (1), 88–121.
Schneider, E., Oppermann, B., Renn, O., 1998. Implementing structured participation for regional level waste management planning. Risk: Health, Saf. Environ. 9, 379–395.
Schweizer, P.-J., 2008. Diskursive Risikoregulierung – Diskurstheorien im Vergleich. Nomos, Baden-Baden.
Schweizer, P.-J.-, Renn, O., Köck, W., Bovet, J., Benighaus, C., Scheel, O., et al., 2016. Public participation for infrastructure planning in the context of the German "Energiewende". Uti. Policy 43, 206–209.
Sclove, R., 1995. Democracy and Technology. Guilford Press, New York.
Science Advise for Policy by European Academies (SAPEA), 2019. Making Sense of Science for Policy under Conditions of Uncertainty and Complexity. Report to the EU Commission (2019), SAPEA, Berlin. Available from: https://doi.org/10.26356/MASOS.
Shrader-Frechette, K.S., 1991. Risk and Rationality: Philosophical Foundations for Populist Reforms. University of California Press, Berkeley and Los Angeles.
Skillington, T., 1997. Politics and the struggle to define: a discourse analysis of the framing strategies of competing actors in a "new" participatory forum. Br. J. Soc. 48 (3), 493–513.
Stern, P.C., Fineberg, V., 1996. Understanding Risk: Informing Decisions in a Democratic Society. National Research Council, Committee on Risk Characterization, National Academy Press, Washington, DC.
Stewart, J., Kendall, E., Coote, A., 1994. Citizen Juries. Institute for Public Research, London.
Stirling, A., 2004. Opening up or closing down: analysis, participation and power in the social appraisal of technology. In: Leach, M., Scoones, I., Wynne, B. (Eds.), Science, Citizenship and Globalisation. Zed, London, pp. 218–231.
Stoll-Kleemann, S., Welp, M., 2006. Stakeholder Dialogues in Natural Resources Management: Theory and Practice. Springer, Heidelberg and Berlin.
Suchman, M.C., 1995. Managing legitimacy: strategic and institutional approaches. Acad. Manage. Rev. 20 (3), 571–610.
Susskind, L.E., Fields, P., 1996. Dealing With an Angry Public: The Mutual Gains Approach to Resolving Disputes. The Free Press, New York.
Sutton, A.J., Jones, D.R., Abrams, K.R., Sheldon, T.A., Song, F., 2000. Methods for Meta-Analysis in Medical Research. John Wiley, London.
Sunstein, C., 2002. Risk and Reason. Cambridge University Press, Cambridge.

Tang, S.Y., Tang, C., Wing-Hung Lo, C., 2005. Public participation and environmental impact assessment in Mainland China and Taiwan: political foundations of environmental management. J. Dev. Stud. 41 (1), 1–32.

Tuler, S., Webler, T., 1995. Process evaluation for discursive decision making in environmental and risk policy. Hum. Ecol. Rev. 2, 62–74.

Tuler, S., Webler, T., 1999. Designing an analytic deliberative process for environmental health policy making in the US Nuclear Weapons Complex. Risk: Health, Saf. Environ. 10 (65), 65–87.

US EPA (Environmental Protection Agency), 1995. Use of Alternative Dispute Resolution in Enforcement Actions. Environment Reporter, 2 June, pp. 301–304.

US EPA (Environmental Protection Agency)/SAB (Science Advisory Board), 2001. Improved Science-Based Environmental Stakeholder Processes. EPA-SAB-EC-COM-01-006, EPA Science Advisory Board, Washington, DC.

US-National Research Council of the National Academies, 2008. Public Participation in Environmental Assessment and Decision Making. The National Academies Press, Washington, DC.

van Asselt, M.B.A., 2000. Perspectives on Uncertainty and Risk. Kluwer, Dordrecht and Boston.

van den Hove, S., 2007. A rationale for science–policy interface. Futures 39 (7), 3–22.

Vari, A., 1995. Citizens advisory committee as a model for public participation: a multiple-criteria evaluation. In: Renn, O., Webler, T., Wiedemann, P. (Eds.), Fairness and Competence in Citizen Participation: Evaluating New Models for Environmental Discourse. Kluwer, Dordrecht and Boston, pp. 103–116.

Vorwerk, V., Kämper, E., 1997. Evaluation der 3. Phase des Bürgerbeteiligungsverfahrens in der Region Nordschwarzwald. Centre of Technology Assessment, Stuttgart, Working Report No 70.

von Winterfeld, D., Edwards, W., 1986. Decision Analysis and Behavioral Research. Cambridge University Press, Cambridge, UK.

Warren, M.E., 1993. Can participatory democracy produce better selves? Psychological dimensions of Habermas discursive model of democracy. Polit. Psychol. 14, 209–234.

Warren, M.E., 2002. Deliberative democracy. In: Carter, A., Stokes, G. (Eds.), Democratic Theory Today: Challenges for the 21st Century. State University of New York Press, Albany.

Wassermann, S., Schulz, M., Scheer, D., 2011. Linking public acceptance with expert knowledge on CO_2 storage: outcomes of a Delphi approach. Energy Procedia 4, 6353–6359.

WBGU (German Advisory Council on Global Change), 2000. World in Transition: Strategies for Managing Global Environmental Risks. Springer, Heidelberg and New York, Annual Report.

Weber, M., 1972. Wirtschaft und Gesellschaft, fifth ed. Kröner, Tübingen (originally published in 1922).

Webler, T., 1995. "Right" discourse in citizen participation: an evaluative yardstick. In: Renn, O., Webler, T., Wiedemann, P. (Eds.), Fairness and Competence in Citizen Participation. Evaluating New Models for Environmental Discourse. Kluwer, Dordrecht and Boston, pp. 35–86.

Webler, T., 1999. The craft and theory of public participation: a dialectical process. Risk Res. 2 (1), 55–71.

Webler, T., 2011. Why risk communicators should care about the fairness and competence of their public engagement process. In: Arvai, J., Rivers, L. (Eds.), Effective Risk Communication. Routledge (Earthscan), Washington, D.C.

Webler, T., Levine, D., Rakel, H., Renn, O., 1991. The group Delphi: a novel attempt at reducing uncertainty. Technol. Forecast. Soc. Change 39 (3), 253–263.

Webler, T., Rakel, H., Renn, O., Johnson, B., 1995. Eliciting and classifying concerns: a methodological critique. Risk Anal. 15 (3), 421–436. June.

Webler, T., Tuler, S., Kruger, R., 2001. What is a good public participation process? Environ. Manage. 27, 435–450.

Wynne, B., 2002. Risk and environment as legitimatory discourses of technology: reflexivity inside out? Curr. Soc. 50 (30), 459–477.

Yosie, T.F., Herbst, T.D., 1998. Managing and communicating stakeholder-based decision making. Hum. Ecol. Risk Assess. 4, 643–646.

CHAPTER

4

Energy transition and civic engagement

Jörg Radtke[1], *Emily Drewing*[1], *Eva Eichenauer*[2], *Lars Holstenkamp*[3], *Jan-Hendrik Kamlage*[4], *Franziska Mey*[5], *Jan Warode*[4] and *Jana Wegener*[4]

[1]Department for Political Science, University of Siegen, Siegen, Germany [2]Research Department Institutional Change and Regional Public Goods, Leibniz Institute for Research on Society and Space (IRS), Erkner, Germany [3]Leuphana University of Lüneburg, Institute of Finance and Accounting, Lüneburg, Germany [4]Kulturwissenschaftliches Institut Essen (KWI), Institute for Advanced Study in the Humanities, Essen, Germany [5]University of Technology Sydney, Institute for Sustainable Futures, Ultimo, NSW, Australia

OUTLINE

4.1 Introduction: framing civic engagement in energy transitions

The German *energy transition* mobilizes civic engagement in society, paradoxically in two rather opposite directions. While self-organized groups and associations run their own renewable energy plants and actively advocate climate protection, other citizens' initiatives campaign against wind farms, grid expansion, and biomass plants. Thus, civic

The Role of Public Participation in Energy Transitions
DOI: https://doi.org/10.1016/B978-0-12-819515-4.00004-0

engagement in the energy transition occurs in various shapes, encompassing national to local levels and bottom-up to top-down initiatives.

Civic engagement is a catch-all term for a multitude of different ways in which individuals and groups influence the community, political decision-making, and other public processes (Ekman and Amnå, 2012). These include voting, party memberships, sending letters, protest actions, cultural hackings, memberships in non-governmental associations, energy cooperatives, interest groups, as well as the attendance of structured participation processes—to name only a few. Thus, the energy transition fosters the emerging and diffusion of different forms of civic engagement. Along with the rise of renewable energies and related infrastructures, manifold initiatives and actions emerged (Radtke et al., 2018), such as participation in bioenergy villages, energy cooperatives, and bottom-up initiatives like urban gardening or car sharing. Apart from this civic engagement in support of the energy transition, opposition and resistance against energy infrastructures are also quite vocal.

As a consequence, we witness more professionalized, individualized, and even creative forms of collective action and participation (van Deth and Maloney, 2012). As a special characteristic of civic engagement in the context of the *energy transition*, a material dimension is emphasized as tangible energy plants are a key element of community energy initiatives (Marres, 2012; Ryghaug et al., 2018). This could change civic engagement as it becomes less focused on a meta-idea and refers to a very specific project-oriented activity, such as the construction of a photovoltaic system. Albeit flanked by social events and meetings, such initiatives prioritize concrete action. Since the beginning of strong civil efforts to foster the energy transition in Germany, a trend toward professionalization can be observed to this day.

From a scientific demand for structuring civic engagement, it is crucial to analyze its character, which could either be *invented participation*, that is, bottom-up grassroots activities, or its top-down counterpart *invited participation*, in which mode, content, and scope of participation are predetermined by incumbents (project developers, political actors, etc.) (Cornwall, 2008). It is frequently argued that invited participation limits or even prevents grass root civic engagement as well as serving as a tool to strengthen dominant discourses on energy transitions rather than opening up the arena for alternative ideas and concepts (Cuppen, 2018). Another focus lays on the concept of social capital, which, as generated through multiple forms of civic engagement, is crucial for a vital democracy (Putnam, 2000).

The public debate on civic engagement and participation within the *energy transition* is highly polarized. Often, the opposition to wind turbines and grid expansion is characterized as egoistic and destructive. Stereotypes like Not in my Backyard (NIMBYs) and "Wutbürger" (enraged citizens) who are hampering the *energy transition* shaped the debate in the past years. In contrast, researchers and politicians euphorically described activism and civic engagement in favor of the transformation as constructive and oriented toward the common good. From their perspective, idealized figures like *change agents* and *pioneers of change* heroically represent the forefront of a dynamic transition toward a sustainable future. The public polarization to connote conflicts as negative and engagement as positive simplifies the issue and does not capture the very nature and complexity of civic engagement. First, defining civic engagement as a continuum with conflict and opposition on one end and engagement on the other obscures the complex issue of engagement in the transformation process. Second, there is no distinct classification whether one or the other pole is associated with a positive or a negative connotation. For example, the nuclear phase-out as a central milestone of the *energy transition* emerged from broad civil resistance and public pressure to overcome this technology and its consequences.

To summarize, civic engagement plays an outstanding role in the energy transition. It is a unique opportunity for individuals and groups to shape the energy transition through their own actions. As not all citizens have the same ideas regarding this matter, conflicts result expectably when different views and opinions collide. In the following, three cases of citizen engagement in the energy transition—community energy and critical citizens' initiatives—will be characterized, and their specific forms of civic engagement will be elaborated.

4.2 Community energy and civic engagement

In the following, community energy is presented as a specific form of civic engagement in the German (renewable) energy sector, commonly termed Bürgerenergie. This definition differs from the prevalent understandings of "community energy," which also contain forms of civic engagement that are not initiated and managed in a bottom-up process. Moreover, there is a plethora of alternative terms such as "civic energy" or "local energy" used in the literature.

This contribution focuses upon equity investments with active participation by shareholders as opposed to (1) other types of financial participation, for example, savings certificates, bearer bonds, subordinated debt, or participation certificates (Holstenkamp, 2019), and (2) "invited participation" in the form of top-down initiated and managed cooperatives or limited liability partnerships. As this latter distinction suggests, it may be difficult to differentiate between active participation and passive investment in specific cases.

Moreover, the vagueness of the term community represents a recurring challenge for researchers. Walker identified six ways in which the notion of community is used in this context "to distinguish an actor, a scale of activity, a spatial setting, a form of network, and a type of process through which carbon reduction objectives can be implemented" (Walker, 2011). This range is usually reduced to two principal interpretations: a "community of location" and a "community of interest" (Walker and Devine-Wright, 2008). Currently, there are more than 1000 energy cooperatives, although not all of them can be classified as local CRE (Kahla et al., 2017).

Most CRE initiatives rely on voluntary work, while some have established (semi-)professional structures. With shrinking feed-in tariffs (FiTs) and the change from FiTs to auctions for larger installations, CRE initiatives have to find new business models, which tend to be more complex. This, in turn, pressures initiatives toward professionalization and establishing cooperations—among CRE initiatives as well as with new or established players in the sector.

CRE is a multifaceted approach to civic engagement, which enables the participation of a broad range of individual and collective actors in energy-related activities. As one of its key characteristics, the actor diversity stands in clear contrast to the centralized monopolistic or oligopolistic configuration of the incumbent energy system while it also differs from large-scale commercial developments and industrial renewable energy (RE) players.

So, who are those actors? Both in the past (anti-nuclear movement) and today (climate and renewable energy action), social movements play a key role in mobilizing CRE initiatives (Hess, 2018). In their most basic form, the traditional "end of wire" consumers have transformed into energy citizens or prosumers, that is, market participants who consume (buy) and generate (sell) clean electricity. Beyond this, the new decentralized technologies also encouraged new market participants to address collective and commercial demands in novel ways. In particular, community-owned energy initiatives such as energy cooperatives pose an opportunity to engage in the transformation of the energy system for actors such as individuals, local governments, and local businesses.

In particular, locally focused forms of CRE convey the notion of civic engagement, offering an opportunity to collectively participate in tangible, project-oriented activities. This is a crucial impetus for many community actors. Driven by a wide range of motivations—from obvious environmental benefits of decarbonizing the energy system to a multitude of social and economic benefits—CRE projects represent a way to engage in the increasingly complex energy systems transformation. Indeed, a common denominator among scholars is that involving people in RE development is a promising approach to raise awareness about and support for progressive RE policy, to build social capital as well as to maintain social legitimacy for RE technologies (Cass et al., 2010).

However, potential members' capability to join and participate (financially) is decisively determined by their disposable wealth or income. Survey results show that male, wealthier, and better-educated people dominate community energy (Radtke, 2016). We suppose that the socio-economic composition of members differs between various types of community energy.

Holstenkamp et al. (forthcoming) link CRE to "energy democracy" and civic engagement, with CRE actors seeking to create alternatives to the dominant energy agenda as well as reclaiming and restructuring the energy sector. Since CRE contributes to the notion of "material democracy" and offers the opportunity to participate in technical equipment as an "extra dimension to political theory debates on democratic governance" (Veelen and van der Horst, 2018: 25ff), it complements existing theories on civic engagement. Ultimately, community energy draws its legitimacy from its (supposed) democratic, participatory, and open character.

However, Hicks and Ison (2018) emphasize that CRE projects reveal "divergent levels of [community] engagement differentiated by the diversity of methods used to facilitate one-way and two-way interaction" (p. 530), moving along a spectrum of involvement in renewable energy. Hence, not every community energy project provides the desired policy outcomes, delivers on community expectations, and conveys community-building motivations behind CRE development. Furthermore, a dilemma can be observed in many cases: while there is a great interest of members in active participation and collaboration, not enough opportunities can be created as (particularly volunteering) project leaders reach their limits (Radtke, 2016). Member activities are a lived reality, but to a lesser extent as one might expect and unequally distributed between various community energy projects. In smaller cooperatives, members play a more active part in terms of collaboration, co-design, and proposing new ideas.

Their participatory and inclusive form of engaging communities in RE deployment can address both key dimensions of energy justice: the distributional and procedural dimensions (Bulkeley and Fuller, 2012). While distributive justice refers to the way costs and benefits are shared, procedural justice is concerned with the decision-making process (e.g., planning and implementation of renewable energy systems) and how inclusive, participatory, and fair it is perceived to be and how it is handled by all stakeholders involved (Gross, 2007). While realistic notions of engagement must be secured (Hoffman and High-Pippert, 2010), community energy has demonstrated its ability to enhance and broaden democratic practices by introducing new actors and perspectives, as well as the nature and quality of dialogues held with various publics (MacArthur, 2016).

In conclusion, CRE is a particular form of civic engagement and can foster societal compromises and consensus in the energy transition. Yet, to harness the full potential of CRE activities, policy support mechanisms have to meet community expectations and help to deliver community motivations behind CRE development.

4.3 Civic engagement and local opposition against wind energy

General support for renewable energy has been high ever since national polls on that matter started in North America and Europe in the 1970s (Walker, 1995). German national polls are no exception to this. But despite the ever-confirming rates of around 90% of respondents agreeing with the goals of the energy transition or approving the expansion of renewable energy infrastructure, the various realizations of projects do not attract undisputed support. Rather, the opposite is true (Walker, 1995; Batel and Devine-Wright, 2015; Reusswig et al., 2016). Especially when it comes to the deployment of wind turbines, the opposition is rising. Local opponents are more and more embedded in and backed by a network of regional or national anti-wind groups and are becoming increasingly organized and professionalized (Reusswig et al., 2016). Thereby, they gain the potential to not only stop a particular project but also to significantly slow down the energy transition in general (Renn and Marshall, 2016). Such headwinds can be observed at many information or participatory events conducted in preparation of wind power projects. The atmosphere in frequently overcrowded assembly halls is often tense, sometimes outright hostile.

Protests predominantly arise in two stages: first, during formal participation processes, when suitable areas for wind park development are designated by the responsible regional planning authorities. Second, opposition comes up when communal licensing authorities grant permission to build a certain wind park project. Common arguments address the issues of landscape aesthetics, potential health impacts, loss of property value, or negative consequences for endangered species. A look beneath the surface of these points of critique reveals that issues of distributional and procedural justice are the most important for local residents. Correspondingly, financial participation is seen as one way to curb protests and to create regional added value. While this might hold true for communal revenues gained through taxes, financial compensation of residents fails to appease active opponents of wind energy (Eichenauer, 2018; Johansen and Emborg, 2018; Lienhoop, 2018). Instead, it is actually perceived as a form of bribery or "hush-money" — even by supporters of wind energy. Hence, when financial compensation is offered in the wrong context, it can even worsen the local conflict (Walker, 1995; Devine-Wright, 2011; Reusswig et al., 2016).

Issues of procedural justice, such as access to unbiased information and proper representation of interests, have been addressed in research regarding the acceptance of renewable energies during the last decade (e.g., Walker, 1995; Bell, Gray, and Haggett, 2005; Gross, 2007; Wolsink, 2007; Lienhoop, 2018). Perceived procedural injustice might even motivate the former supporters of local wind energy to actively engage in anti-wind protests (Wolsink, 2007). Discontent with the participation process, in general, may relate to a variety of issues (Eichenauer, 2018). For example, access to a formal participation process can be viewed as neglect of neighboring communities' residents. Though equally affected, they are not entitled to any information on the community's siting decisions, let alone being able to raise objections. If objections against planning decisions are indeed raised, activists complain that their objections are not taken seriously. While it is also true that the vast majority of the objections is not legally relevant (e.g., matters of visual impact or expected loss of property value), it leaves concerned residents flabbergasted, if—as happened during a regional planning process in Northeast Germany—all of nearly 90,000 objections that had been turned in were rejected. Mistrust in information given at public events or in preparation of the building permits is also frequently criticized. Though cautious and compassionate risk communication is essential in the forerun of large-scale infrastructure projects, the residents report that their concerns, be it infrasound or the thread of potential forest fires caused by burning wind turbines, were marginalized or sometimes even ridiculed. Also, major doubts are raised concerning the objectivity and scientific soundness of avifaunistics or other required expertise that are commissioned by the project developers. This leads to an environment in which mistrust in information and the involved actors grows while trust in the institutions of a democratic *constitutional state* declines.

Adding to this, (and despite the plurality of arguments of critics as shown above), a common explanation for opposition against large-scale infrastructure projects such as wind turbines is still prevalent within the discourse in the shape of the so-called NIMBYs (e.g., Engels et al., 2013): while people generally agree to the expansion of renewable energies — as shown by the overwhelming 90% majority of the German public that supports the energy transition - they are not willing to accept them in their direct vicinity, which appears evident with regard to the increase in protests against local projects or development plans (hence, NIMBY). However, this conceptualization is not only empirically misleading since it offers a simplified explanation for a very complex social phenomenon, but also discredits the residents and their concerns, delegitimizes their opposition as being irrational, and defames the protesters as acting out of mere self-interest and ignoring a common good and goal (which is the realization of the energy transition through the deployment of a certain wind power project) (Wolsink, 2007; Devine-Wright, 2011).

While there is no doubt that debates on zoning and licensing are often stirred by emotions, fear and invalid or even false arguments, the non-recognition of well-founded and valid arguments as well as defaming them as mere selfish NIMBYs might lead to sociopolitical issues that transgress local energy political disputes. In the light of growing support of populist parties that do not only challenge liberal democracies across Europe but also promote climate skepticism and aim at stopping or even reversing the energy transition, dismissing citizens' genuine concerns offhandedly appears particularly negligent.

While anti-wind activists are not necessarily climate skeptics or opponents of the energy transition and its general goals (Eichenauer, 2018), the feeling of being constantly sidelined within so-called participation processes which are supposed to add democratic legitimization to the energy transition might lead to the perception of the energy transition as an elitist undertaking that does not take the needs and sorrows of the affected population into consideration. This is a door opener for populist parties that challenge not only the transformation of the energy system but also the democratic political system in general.

4.4 Grid expansion and civic engagement

Characterized as the *energy transition's* "bottleneck" (Steinbach, 2013), the expansion of the transmission grid constitutes a decisive moment in the progress of the German energy transition as it fosters the system and market integration of the rapidly growing renewable energies. The rise of the renewables, however, marks a paradigm shift for the whole energy system, namely the transition from a rather centralized structure of energy production by an oligopoly toward a more decentralized production system including a variety of actors. Therefore, the extension of the grid is itself a part of a profound and highly contested societal negotiation process on the future energy system's degree of centralization.

Since wind energy production has expanded especially in the windier north of Germany and the electricity demand by industrial consumers is particularly high in the south, the existing transmission lines lack the capacity to cover the rising power volume. Consequently, new transmission lines have been planned. According to the German Federal Ministry for Economic Affairs and Energy, around 5900 kilometers of transmission grid needs to be built in the near future, while only 250 kilometers have been constructed by the end of 2018 (Bundesnetzagentur, 2019). The four transmission system operators estimate the expansion cost to amount to 52-billion Euros until 2030 (50Hertz Transmission, 2019: 143ff).

Although the German population generally supports the *energy transition* (Setton et al., 2017), many people engage in protests as risks and burdens manifest locally. The mere scale of the required infrastructure extension implies far-reaching interferences with people's everyday lives and the environment. Interventions of this dimension change the perception of landscape and turn familiar spaces into *energy landscapes* (Apostol et al., 2016). Kühne and Weber show that citizen protest against grid expansion is often based on worries about "disfigurement of the landscape", "destruction of nature", and "loss of home environment" (Kühne and Weber, 2018). Consequently, the grid extension's success depends decisively on the willingness of directly affected local populations to bear potential risks and negative effects for the community.

As the number of local and regional conflicts constantly increases, a network of citizen initiatives has emerged (Neukirch, 2016) and newly founded umbrella organizations and protest groups indicate a dynamic process of professionalization. Internet platforms and social media networks offer additional spaces and opportunities for the exchange of information and findings, common campaigning, and other coordinated actions.

Albeit the relevant literature on resistance against grid expansion mainly consists of case studies that vary in scope and methodology, common aspects can be identified: first of all, most protests are carried out by coalitions of local actors (Neukirch, 2016). These coalitions include directly affected citizens, environmental activists, mayors, local and federal politicians, and the local media (Kamlage et al., 2018). Citizens' initiatives are becoming increasingly important in the context of power grid extension (Weber et al., 2017). They play a major role as focal points of information and communication flow for locals (Schweizer-Ries et al., 2010), organize events, express local interests, and even negotiate with net operators, planning administrators, and politicians (Neukirch, 2016). Contrary to the common belief, most protests are not merely driven by NIMBY motives. According to Weber et al. (2017), local opposition is primarily rooted in concerns over the alienation of landscapes and home environments accompanied by the fear of health impairment through electromagnetic radiation. Secondary issues are nature conservation and concerns about local economic development, such as negative impacts on property values and local tourism. The opposition is often expressed in demonstrations and public debates, resolutions and legal actions against the measures disputed. Apart from the protest carried out by actors who live in the concerned region, there is a more general type of opposition challenging the underlying concept of grid extension, criticizing the decision-making process and arguing that a more decentralized energy system would not require the planned power lines (Neukirch, 2016).

To sum up, the power grid expansion in Germany is challenged predominantly by local conflicts driven by coalitions of actors, which delay the progress of the energy transformation. We suggest to understand conflicts and engagement as catalysts of social transformation and to note that both phenomena include constructive and functional characteristics. However, politicians and researchers suggest public participation as a remedy to tackle the lack of acceptance and deadlocks of the expansion of the grid. Recent empirical studies support this argument: directly affected citizens express the need for direct and impactful face-to-face participation in grid planning processes (Schweizer-Ries et al., 2010). They demand transparent, early-on, comprehensive, and balanced information about infrastructure planning (Hübner and Hahn, 2013; Bertsch et al., 2016). Structured forms of dialogue-based participation enhance the level of information and establish trustful relations among involved actors. Face-to-face participation supports shared views on problems and encourages a more rational debate based on reasoning (Schweizer et al., 2016). If structured participatory processes are professionally designed in line with commonly shared standards of fairness and justice, they may raise the democratic legitimacy of the grid expansion (Hildebrand et al., 2012; Kamlage et al., 2014). Nevertheless, participation is "no panacea for enhanced social acceptance" (Schweizer and Bovet, 2016: 64ff) and its success highly depends on both the quality of the process design and the respective circumstances.

4.5 Conclusion

The examples of community energy and citizens' initiatives show that there are many opportunities for citizens and organized interests to engage in the energy transition. Their civic engagement can roughly be differentiated according to whether it is strongly shaped by bottom-up action (citizens start an action group together) or rather a result of an invitation (offer to participate in predefined settings). There is no consistent definition of civic engagement in general, and therefore not in the context of the energy transition in particular. Basically, it emphasizes the self-initiative of citizens, it is based upon voluntary and unpaid activity and it finds its limits where, with increasing formalization and institutionalization, the professional level of an organization is achieved.

Two basic types of engagement in the energy transition can be distinguished:

- An association of citizens in a community who are committed to climate protection and aim to carry out a project for this purpose. Community energy initiatives are in a sense a response to less action–oriented climate protection discussion groups. There are, however, two types of community energy: first, a bottom-up initiative that can be considered a form of civic engagement as a group of citizens establishes an energy cooperative in its local community. Second, a more professional version of the former, as an energy supply company (frequently non-profit based) or local bank, offers financial participation—this form would be incompatible with the concept of civic engagement. However, the lines between these varieties blur, indicating that distinctions between invented and invited participation are fuzzy.
- In addition, there are numerous citizens' initiatives, which take a critical stance toward the energy transition. Their goal is not necessarily to prevent a specific implementation such as the construction of a wind turbine. Mainly, they demand more options to deliberate. Their actions can clearly be conceptualized as a form of civic engagement as well as political participation. The variety of initiatives in terms of their motivation, membership structure, goals, their actual actions as well as their interaction is vast. Again, invented participation can turn into an invited form. Similar to community energy, these initiatives tend to focus on concrete projects and thrive on their dynamics, to which they usually respond or intervene against. The process character is very typical of citizens' initiatives, which often focus on one goal (e.g., strong support or prevention of a certain measure).

To conclude, citizens' initiatives either opposing the *energy transition* or realizing their own projects to further its progress are a reflection of society as they mirror an increasing polarization. However, this picture is deceptive, as communities are more heterogeneous than they may appear. The citizens' initiatives are strongly attached to their spatial as well as mental spheres and often seek to influence a public discourse decisively. Therefore, they are an example of a high politicization. The degree of binding social capital is usually very high in the case of citizens' initiatives (internal cohesion), but concerning the energy transition as well as local democracies, external bridging interactions are more important. However, the high degree of politicization tends to prevent compromises and consensus building. Thus, intermediary actors play a decisive role, as well as strategies that emphasize connecting factors (Lacey-Barnacle and Bird, 2018). The influence and power of citizen movements should ultimately not be underestimated. According to Ulrich Beck, they facilitate the shaping of sub-politics and through agenda setting, they can significantly influence the design of policies.

References

50Hertz Transmission, Amprion, TenneT TSO, and TransnetBW, 2019. Netzentwicklungsplan Strom 2030. Erster Entwurf der Übertragungsnetzbetreiber. Berlin et al.

Apostol, D., Palmer, J., Pasqualetti, M., Smardon, R., Sullivan, R., 2016. The Renewable Energy Landscape: Preserving Scenic Values in Our Sustainable Future. Taylor and Francis Ltd, London, New York.

Bertsch, V., Hall, M., Weinhardt, C., Fichtner, W., 2016. Public acceptance and preferences related to renewable energy and grid expansion policy: empirical insights for Germany. Energy 114, 465–477.

Bulkeley, H., Fuller, S., 2012. Low carbon communities and social justice. Joseph Rowntree Foundation Viewpoint. <http://www.jrf.org.uk/sites/files/jrf/low-carbon-communities-summary.pdf>.

Bundesnetzagentur, 2019. Netzausbau—Leitungsvorhaben. <https://www.netzausbau.de/leitungsvorhaben/de.html;jsessionid = B9D5F50705C93D49F6B8>.

Cass, N., Walker, G., Devine-Wright, P., 2010. Good neighbours, public relations and bribes: the politics and perceptions of community benefit provision in renewable energy development in the UK. J. Environ. Pol. Plann. 12 (3), 255–275.

van Deth, J.W., Maloney W.A., 2012. New participatory dimensions in civil society. Professionalization and individualized collective action. Routledge ECPR Studies in European Political Science, Routledge, Abingdon.

Devine-Wright, P., 2011. From backyards to places: public engagement and the emplacement of renewable energy technologies. In: Devine-Wright, Patrick (Ed.), *Renewable Energy and the Public:* From NIMBY *to Participation.* Earthscan, London, pp. 57–70.

Eichenauer, E., 2018. Energiekonflikte – Proteste gegen Windkraftanlagen als Spiegel demokratischer Defizite: Politikwissenschaftliche Perspektiven. In: Radtke, Jörg, Kersting, Norbert (Eds.), *Energiewende:* Politikwissenschaftliche Perspektiven. Springer VS, Wiesbaden, pp. 315–341.

Ekman, J., Amnå, E., 2012. Political participation and civic engagement: towards a new typology. Hum. Aff. 22 (3), 283–300.

Engels, A., Hüther, O., Schäfer, M., Held, H., 2013. Public climate-change skepticism, energy preferences and political participation. Global Environ. Chang. 23 (5), 1018–1027.

Gross, C., 2007. Community perspectives of wind energy in Australia: the application of a justice and community fairness framework to increase social acceptance. Energ. Policy 35 (5), 2727–2736.

Hess, D.J., 2018. Energy democracy and social movements: a multi-coalition perspective on the politics of sustainability transitions. Energ. Res. Soc. Sci. 40, 177–189.

Hicks, J., Ison, N., 2018. An exploration of the boundaries of 'community' in community renewable energy projects: Navigating between motivations and context. Energ. Policy 113, 523–534 (February 1, 2018).

Hildebrand, J., Rau, I., Schweizer-Ries, P., 2012. Die Bedeutung dezentraler Beteiligungsprozesse für die Akzeptanz des Ausbaus erneuerbarer Energien. Eine Umweltpsychologische Betrachtung. Informationen zur Raumentwicklung 9–10, 491–502.

Hoffman, S.M., High-Pippert, A., 2010. From private lives to collective action: recruitment and participation incentives for a community energy program. Energ. Policy 38 (12), 7567–7574.

Holstenkamp, L., 2019. Financing consumer (co-)ownership of renewable energy sources. In: Lowitzsch, Jens (Ed.), *Energy Transition: Financing Consumer* Co-Ownership *in Renewables.* Palgrave Macmillan, Cham, pp. 115–138.

Hübner, G., Hahn, C., 2013. Akzeptanz des Stromnetzausbaus in Schleswig-Holstein. Abschlussbericht zum Forschungsprojekt. Deutsche Umwelthilfe e.V. und Bundesministerium für Umwelt, Naturschutz und Reaktorsicherheit. Halle.

Johansen, K., Emborg, J., 2018. Wind farm acceptance for sale? Evidence from the Danish wind farm co-ownership scheme. Energ. Policy 117, 413–422.

Kahla, F., Holstenkamp, L., Müller, J.R., Degenhart, H., 2017. Entwicklung und Stand von Bürgerenergiegesellschaften und Energiegenossenschaften in Deutschland. <http://www.leuphana.de/fileadmin/user_upload/Forschungseinrichtungen/professuren/finanzierung-finanzwirtschaft/files/Arbeitspapiere/wpbl27_BEG-Stand_Entwicklungen.pdf>.

Kamlage, J.-H., Nanz, P., Fleischer, B., 2014. Bürgerbeteiligung und Energiewende: Dialogorientierte Beteiligung im Netzausbau. In: Ekardt, Felix, Grothe, Anja, Hasenclever, Wolf-Dieter, Hauchler, Ingomar, Jänicke, Martin, Kollmann, Karl, Michaelis, Nina V., Nutzinger, Hans G., Rogall, Holger, Scherhorn, Gerhard (Eds.), Viertes Jahrbuch Nachhaltige Ökonomie. Metropolis, Marburg.

Kamlage, J.-H., Richter, I., Nanz, P., 2018. An den Grenzen der Bürgerbeteiligung: Informelle dialogorientierte Bürgerbeteiligung im Netzausbau der Energiewende. In: Holstenkamp, Lars, Radtke, Jörg (Eds.), Handbuch Energiewende und Partizipation. Springer VS, Wiesbaden, pp. 627–642.

Kühne, O., Weber, F., 2018. Conflicts and negotiation processes in the course of power grid extension in Germany. Landsc. Res. 43 (4), 529–541.

Lacey-Barnacle, M., Bird, C.M., 2018. Intermediating energy justice? The role of intermediaries in the civic energy sector in a time of austerity. Appl. Energy 226, 71–81.

Lienhoop, N., 2018. Acceptance of wind energy and the role of financial and procedural participation: an investigation with focus groups and choice experiments. Energ. Policy 118, 97–105.

MacArthur, J.L., 2016. Challenging public engagement: participation, deliberation and power in renewable energy policy. J. Environ. Stud. Sci. 6 (3), 631–640.

Marres, N., 2012. Material Participation: Technology, the Environment and Everyday Publics. Palgrave Macmillan, Houndmills, Basingstoke.

Neukirch, M., 2016. Protests against German electricity grid extension as a new social movement? A journey into the areas of conflict. Energ. Sustain. Soc. 6 (1), 4.

Putnam, R.D., 2000. Bowling Alone: The Collapse and Revival of American Community. Simon and Schuster, New York.

Radtke, J., 2016. Bürgerenergie in Deutschland. Partizipation zwischen Rendite und Gemeinwohl. Schriftenreihe Energiepolitik und Klimaschutz. Springer VS, Wiesbaden.

Radtke, J., Holstenkamp, L., Barnes, J., Renn, O., 2018. Concepts, formats, and methods of participation: theory and practice. In: Holstenkamp, Lars, Radtke, Jörg (Eds.), *Handbuch Energiewende und* Partizipation. Springer VS, Wiesbaden, pp. 21–42.

Renn, O., Marshall, J.P., 2016. Coal, nuclear and renewable energy policies in Germany: From the 1950s to the "Energiewende". Energ. Policy 99, 224–232.

Reusswig, F., Braun, F., Heger, I., Ludewig, T., Eichenauer, E., Lass, W., 2016. Against the wind: local opposition to the German 'Energiewende'. Util. Pol. 41, 214–227.

Ryghaug, M., Skjølsvold, T.M., Heidenreich, S., 2018. Creating energy citizenship through material participation. Soc. Stud. Sci. 48 (2), 283–303.

Schweizer, P.-J., Bovet, J., 2016. The potential of public participation to facilitate infrastructure decision-making: lessons from the German and European legal planning system for electricity grid expansion. Util. Pol. 42 (October), 64–73. Available from: https://doi.org/10.1016/j.jup.2016.06.008.

Schweizer, P.-J., Renn, O., Köck, W., Bovet, J., Benighaus, C., Scheel, O., et al., 2016. Public participation for infrastructure planning in the context of the German "Energiewende". Util. Pol. 43 (December), 206–209.

Schweizer-Ries, P., Zoeller, J., Rau, I., 2010. Akzeptanz Neuer Netze: Die Psychologie der Energiewende. In: Boenigk, N., Franken, M., Simons, K. (Eds.), *Kraftwerke Für Jedermann: Chancen und Herausforderungen einer dezentralen erneuerbaren Energieversorgung. Sammelband* Dezentralität. LokayDruck, Reinheim, 60–63.

Setton, D., Matuschke, I., Renn, O., 2017. Social Sustainability Barometer for the German Energiewende 2017. Core Statements and Summary of the Key Findings. Institute for Advanced Sustainability Studies e.V. Potsdam.

Steinbach, A., 2013. Barriers and solutions for expansion of electricity grids—the German experience. Energ. Policy 63 (December), 224–229.

van Veelen, B., van der Horst, D., 2018. What is energy democracy? Connecting social science energy research and political theory. Energ. Res. Soc. Sci. 46, 19–28.

Walker, G., 1995. Renewable energy and the public. Land Use Policy 12 (1), 49–59.

Walker, G., 2011. The role for "community" in carbon governance. WIRES Clim. Change 2 (5), 777–782.

Walker, G., Devine-Wright, P., 2008. Community renewable energy: what should it mean? Energ. Policy 36 (2), 497–500.

Weber, F., Jenal, C., Rossmeier, A., Kühne, O., 2017. Conflicts around Germany's Energiewende: discourse patterns of citizens' initiatives. Quaestiones Geographicae 36 (4), 117–130.

Wolsink, M., 2007. Planning of renewables schemes: deliberative and fair decision-making on landscape issues instead of reproachful accusations of non-cooperation. Energ. Policy 35 (5), 2692–2704.

CHAPTER

5

From coal to renewables: changing socio-ecological relations of energy in India, Australia, and Germany

Tom Morton[1], *Jonathan Paul Marshall*[1], *Linda Connor*[2], *Devleena Ghosh*[1] *and Katja Müller*[3]

[1]University of Technology, Faculty of Arts and Social Sciences, Sydney, Australia [2]University of Sydney, NSW, Australia [3]Martin-Luther-Universität Halle-Wittenberg, Zentrum für Interdisziplinäre Regionalstudien (ZIRS), Germany

OUTLINE

What we do over the next ten years will determine the future of humanity for the next ten thousand years. —***Prof. Sir David King, former chief scientific advisor, UK Government, BBC News, 10.5.2019.***

In May 2019, announcing an urgent UN Climate Action Summit for September that year, the UN Secretary-General António Guterres called for a moratorium on new coal-fired power plants by 2020 and noted that "if all coal power plants currently under construction go into operation and run until the end of their technical lifetime, emissions will increase by another 150 gigatonnes, jeopardizing our ability to limit global warming by 2°C" (UN News Centre, 2019).

As Guterres stressed, an immediate exit from coal-fired power is the key priority in confronting the climate emergency and achieving the Paris climate goals. We argue in a

The Role of Public Participation in Energy Transitions
DOI: https://doi.org/10.1016/B978-0-12-819515-4.00005-2

forthcoming publication (Goodman et al., 2020) that there is conflicting evidence as to whether or not the world has really reached "peak demand" for coal as an energy source. Projections from industry bodies and corporations, such as the IEA, ExxonMobil, BHP Group, and BP, identify apparently contradictory energy trends. BHP Group predicts that coal will "progressively lose competitiveness to renewables on a new build basis in the developed world and in China" but still be competitive in India and other emerging markets (McKay, 2019). The BP World Energy Outlook states that "renewables are the largest source of energy growth, gaining at an unprecedented rate" but acknowledges that world demand for coal is unlikely to decline unless governments take much more decisive action to curb emissions (Smith, 2019). The IEA remarks that despite the rapid growth of renewables, the pace and scale of energy transition "is not in line with climate targets" (International Energy Agency, 2019). These narratives of a surging global energy transition, and stubbornly intractable demand for fossil fuels, exist side by side.

This chapter reports on three countries, Germany, India, and Australia, which Climate Transparency (2018) claims are unlikely to meet their immediate Paris Agreement targets. Germany *is* decreasing emissions, while Australia and India are increasing. Just because we may need a transition does not mean it will occur, or occur in the manner in which we need it to occur. Consequently, as Kern and Rogge (2017: 13) argue, the pace of energy transitions and whether they can be sped up is a key academic and policy question.

In the following discussion, we understand transition as involving both a process of contesting the meaning, legitimacy, and use of fossil fuels, especially coal, and a process of building a social and political constituency for transition to a relatively decarbonized society, powered by renewable energy. Energy transition is not simply a technocratic process, and it is inevitably bound up with energy governance, which according to Szulecki is at a crossroads, facing a "third industrial revolution" (Szulecki, 2018: 21). Kern and Rogge (2017: 13) point out that while in the past "energy transitions have not been consciously governed," today a wide variety of social groups are actively engaged in attempting to build transition toward low greenhouse gas emission energy systems. The transition opens up possibilities for public input into energy decision-making, design, and implementation, as well as generating new conflicts.

In the following discussion, we analyze the governance of energy transitions from the ground up, drawing on findings from a comparative research project on the contestation of coal mining in Australia, India, and Germany, and a new project on energy transition in the same countries. All three countries are coal dependent, with the state being heavily involved in coal extraction as a developmentalist project; all are parliamentary democracies, although with very different characteristics; and each is engaged in different processes of decarbonization, delay, and energy transition. In each case, the pace of this transition, and its sufficiency, is heavily intertwined in social struggle, issues of political acceptability, and cultural sense-making, as well as relying on technological innovation and fit with existing economic systems. Those in the coal-industrial complex have exerted their influence to inhibit the change and recognition of a need for change in all three countries, but state policies on energy transition vary widely. Germany has committed to an ambitious policy of "full decarbonization of the economy" by mid-century and "the transition to an energy system in which energy supply is almost fully based on renewable energies" (Fabra et al., 2015: 51). India's Draft National Electricity Plan foresees that 57%

of total electricity capacity will come from non-fossil-fuel sources by 2027, although it is expected that coal use will continue to expand (Government of India, 2016; McKay, 2019). Australia has experienced rapid growth in renewable investment in recent years, driven by the private sector, but while industry commentators describe this as "a once-in-a-lifetime change in the energy supply paradigm," the same commentators fear new investment may "fall off a cliff" when the current Renewable Energy Target expires in 2020 (MacDonald-Smith, 2018). Since the demise of the National Energy Guarantee in 2018, Australia has had no coherent long-term national energy policy and investors have been in a state of uncertainty.

Rather than focusing on national policy frameworks—or the lack of them—our research, drawing on local ethnographies conducted over 5 years from 2014 to the present, seeks to understand what the process of energy transition means to those caught up in it, in specific contexts. In our forthcoming book (Goodman et al., 2020), we investigate how the contestation of coal extraction, and recognition of climate change, could be delegitimizing coal, calling into question the social, political, and cultural meanings, which have underpinned its value as a developmental commodity since the Industrial Revolution. Contests over coal are also contests over narrative, and as the Trilateral Group on European Structural Change has noted, creating a "narrative for change" is a necessity if large-scale energy transitions are to be successful (Heinrich Böll Stiftung, 2018).

In these concrete local contexts, the success of transition also involves the recognition of potential problems with renewable energy, the social ways of organizing and installing it, and the building of resistances to the transformation. In all instances, the ways that people are identified or rejected as "stakeholders" exert an important influence on the pathways of transition. Formal processes of identification of stakeholders, consultation, and planning may well exclude poorer and more marginal people who are most likely to be severely hurt by ecological catastrophe. We also investigate the extent to which contestation of new or expanding coal mines is framed in relation to energy transition. The motivations of the individuals and communities involved in these contestations are not necessarily expressed in terms of energy policy, energy democracy, energy transition, or climate action, although in some cases they are. However, the process of contestation may open up new possibilities and social spaces for participation by citizens in the control and preservation of their own lives and lead to new awareness of the forces opposing them. We point to the kinds of support and hindrance available to people: legal, regulatory, civic, and others. We ask whether transition to a sustainable energy future, in the context of climate change (where relevant), provides an "alternative narrative" to (continuous) development through coal burning.

As a part of our attempt to explore how public input and public involvement should be designed, structured, and organized so that it facilitates the transition toward a more sustainable energy future, we make some preliminary comparisons between relations of renewable energy transition and coal contestation. We point to contexts where renewable energy projects are imposed in ways similar to those in which coal mines are imposed, often generating protest, affecting land or water use, encroaching on residential areas, or seeming to maintain existing power structures and patterns of social exclusion.

It cannot necessarily be assumed that the forms of governance emerging through a transition to renewable energy will be successful, beneficial, or more democratic. Similarly, in Kern and Rogge's (2017: 13) terms, not all the actors engaged in "active attempts to govern the transition towards low energy carbon" may continue to support that transition if they feel excluded from the process or disappointed by its particular implementation. Without awareness of this, the process risks are being unsuccessful at huge cost to us all.

5.1 Germany

Our research in Germany focussed initially on the Lausitz coal-mining region in Eastern Germany, and specifically in Kerkwitz, Atterwasch, and Grabko, three villages close to the Polish border. Since 2007, the villagers have been fighting a proposed expansion of the nearby Jänschwalde lignite mine, owned at the time by the Swedish company Vattenfall. Vattenfall had lodged an application to extend the area of the mine by 2000 ha, which, if approved, would have enabled the mining of an additional 200 million tonnes of lignite over 20 years and necessitated the demolition of the three villages.

After the initial shock, local opposition grew quickly. Several citizen initiatives formed to monitor the Jänschwalde-Nord mine-planning process, generate strategies of resistance and promote protest. They initially tried direct democracy, and in 2007, mine opponents collected more than 20,000 signatures required for a popular petition (*Volksinitiative*). The state parliament of Brandenburg was forced to debate the mine extension, but without any tangible result. Over time, the villagers and their supporters in the region settled on a strategy of low-level, repetitive protest, which we have described previously as a strategy of passive and active waiting, or waiting and delaying (Műller and Morton, 2018; Műller, 2019).

Mine opponents intervened in the slow mine approval processes, said to take an average of 6–10 years, whenever possible. For example, in 2011, they objected to the environmental impact assessment and, throughout the planning process, partook in the *Braunkohlenausschuss*, an assembly of delegates from the relevant city councils and civil society, with an advisory function to the state government. The *Braunkohlenausschuss* includes both proponents and opponents of the mine and requires constant negotiations and discussions. The sheer bureaucratic tedium of the planning process, and the uncertainty of its outcome, took a toll on the morale of the villagers, but it also gave them time in which to organize, form alliances, build solidarity, and develop their own rituals of resistance. For example, the *Sternmarsch*, in which mine opponents walk from one village to another, became a ritual every first Sunday in January. The *Dorffest*, the village fête in Atterwasch, enhanced the *Sternmarsch* as another repetitive form of protest, taking place on the last day of every October since 2012.

These local rituals were complemented by the input of the *Lausitzer Klima- und Energiecamp* (Lausitz Climate and Energy Camp), which entered the local protest in 2011. Local actions not only built local constituencies but also lead to two major protest actions, expanding those constituencies and placing the anti-coal mine narrative in a global context. One action was the human chain against coal in 2014, when 7500 protesters linked hands to form a chain from Kerkwitz to Grabice, 8 km away in Poland on the other side of the river Neisse. The second major protest was *Ende Gelände* (closed ground) in May 2016, when protesters occupied part of the Welzow-Sűd open-cut lignite mine and the nearby Schwarze Pumpe power plant, forcing the plant to significantly reduce its operations. *Ende Gelände* was one of the 20 simultaneous protests on six continents organized by "Break Free from Fossil Fuels," a global campaign against fossil fuels (https://breakfree2016.org/).

Both *Ende Gelände* and the human chain protest explicitly linked local protesters with national and transnational climate activists, drawing a clear connection between the local contestation of coal and coal's centrality to climate change. They also demonstrated the need for an energy transition, which the villagers argue that they are already living, through the local installation of solar panels, biogas plants, and small wind turbines. For the residents of these three villages, the energy transition is integrated into the fabric of German rural life and the *Energiewende* provides an alternative narrative to the one of coal as the driving force of the local and nation-wide economy (Morton and Müller, 2016). "Playing for time," while this energy transition gathers momentum, gives their endurance a wider narrative meaning.

This strategy proved successful. In March 2017, EPH, the Czech energy company, which had bought all Vattenfall's coal mines and power plants in the Lausitz, announced that they were abandoning the extension of Jänschwalde-Nord. The villagers had won. We are not suggesting that the villagers' resistance alone lead to this outcome: EPH portrayed its withdrawal as purely economic, but through their various forms of organization, participation, and protest, mine opponents were able to make input into energy decision-making, design, and implementation. They became "relevant stakeholders" in the process, against resistance that would have excluded them.

One of the ironies that emerge from our preliminary fieldwork in Brandenburg, in the region of Niederer Fläming, is that similar strategies of waiting and delaying may be used to oppose renewable energy, albeit in different actor constellations. Local opponents of wind energy do not stage protests with anywhere near the scale and regularity of the anti-coal activists in the Lausitz, but planning processes for wind farms have become highly complex, involving local, regional, and federal governments, and subject to legal challenge. Local activists against (particular) wind farms intervene in the planning and installation process whenever possible. Some of our interviewees formed a party and got into local parliament, becoming better informed and more able to scrutinize wind park planning. Their take on renewable energy is by no means a dismissive one; wind critics opting immediately for coal or nuclear energy are rare, but the number of wind parks being built in close proximity to residential spaces renders issues of the visual appearance of wind turbines in the landscape, sound, and infrasound, or the blinking of warning lights at night important. In practice, wind opponents often take up the (administratively predetermined) narratives of protection of the local environment, the forest, and local fauna, particularly birdlife, to justify their opposition. This, as well as administrative requirements, extends the approval process for a wind park, so it can take about 3 years from application to decision. Including the planning and construction period makes a likely total of 6 years, or more, for building; a single wind turbine can take as long as a (quick) approval procedure for an open-cut coal mine.

Local residents in Niederer Fläming—as in other areas with a high density of wind farms—bear some of the costs of renewable energy production, through the impacts outlined above, and through slightly higher electricity prices. While some individuals leasing their land for wind farms also profit financially, national and state governments have installed mediation agencies and redistribution mechanisms to cushion the distributive effects of costs and profits from renewable installation. These procedures allow whole villages rather than single landowners to benefit financially from wind farms, and tensions between "stakeholders" can be discussed with neutral experts. However, anything worth naming "energy democracy" still seems far away, especially with legislative changes, such as the replacement of feed-in tariffs with reverse auctions, which favor larger players over energy cooperatives (*Bürgerenergie*) and increase corporate control over the grid and energy production.

5.2 India

Initial fieldwork in India spanned 3 years (2014–16) in the central Indian state of Chhattisgarh. Research focused on three villages in Sarguja district, two of which, Salhi and Ghatbarra, were particularly affected by the Parsa East Kete Basan mine and its extensions, and the third, Madanpur, was the site of protest movements against this expansion. Due to intense government surveillance, our trips were necessarily short in duration and planned around particularly important events (Ghosh, 2018). We focused on the ways in which villagers used legal and democratic processes to assert their rights and sought to understand the complex, sometimes contradictory, motivations, and actions of the various would-be "stakeholders," some of whom, initially, were not opposed to the mines.

Mining in Chhattisgarh is both enabled and made complicated because much of the coal lies beneath the pristine, dense, and contiguous tracts of forests (Greenpeace India, 2012) occupying over 40% of the state. These forests contain perennial water sources, rare plants, and wildlife species, including elephants and leopards. About a third of Chhattisgarh's population consists of Indigenous peoples (Scheduled Tribes or *Adivasis*); they make up about 10% of the *Adivasis* in India (Ministry of Tribal Affairs, 2013). They are mostly forest dwellers depending on the forests for their livelihood. In 2009, issues over land, livelihood, and resources became so fraught that trade unions, community groups, and other progressive parties formed an alliance called the *Chhattisgarh Bachao Andolan* (Save Chhattisgarh Movement). *Adivasis* drew inspiration and narrative for their protests from their special status in the Indian constitution; they consider that their core rights, autonomy, and dignity are set out in Article 21 and under Schedules V and VI. The struggle around their right to *jal, jangal, jameen* or "water, forests, land," as opposed to the developmental goals of coal power, has become a test of the resilience of the Indian Constitution and its guarantees for the protection of minorities.

Opposition to the mines has mostly been mediated through the Panchayat Extension to Scheduled Areas Act (PESA 1996) and the Scheduled Tribes and Other Traditional Forest Dwellers (Recognition of Forest Rights) Act (FRA 2006), two landmark pieces of federal legislation, which attempted to ameliorate the continuing injustices suffered by *Adivasis* and other forest-inhabiting communities since the time that the forests were "reserved" under colonial rule. PESA mandated consultation with *Gram Sabhas* (Village Assemblies) or *Panchayats* before land recognized under the fifth Schedule of the Constitution could be acquired or alienated for development projects (Lahiri-Dutt et al., 2012).

The FRA was co-written by *Adivasi* activists, and the Preamble speaks of righting the "historical injustice" experienced by Adivasis (Ahmad, 2014). It put in place a clear legal mechanism for recognizing rights at both individual and community levels for forest dwelling communities, including forest workers, who had lived in a designated forest area for 75 years or three generations. It recognizes and vests secure community tenure on "community forest resources" in *Gram Sabhas* or village assemblies (Ministry of Tribal Affairs, 2013). However, state authorities devised means of sidestepping these protections so that these Acts have become the focus of local and national struggles in new and potentially transformative ways. For example, in January 2015, representatives of 20 *Gram Sabhas* in Chhattisgarh met the ministers of Environment and Tribal Affairs to demand that the government stop the auction of coal-mining licenses in their districts, to either the private or public sector, as they would not consent to mining. As elected representatives of local village councils (*Gram Panchayat*), they were exercising their constitutional mandate as articulated through the PESA and the FRA.

The villagers have the following three sets of grievances: that due compensation is not received for land acquired for mining, nor is rehabilitation mandatory; that land acquisition is often unsafe, illegal, or coerced, and contrary to the wishes of the *Gram Sabhas*; and that socio-ecological change caused by the mines, such as pollution of the waterways and impact on wildlife, is not ameliorated. Their strategy of opposition is through their constitutional rights: withholding consent at the *Gram Sabha* level, applying for individual and community forest rights, contesting environmental and other violations before the National Green Tribunal, and taking cases of malfeasance and misappropriation to the courts. In addition, there are civil actions, protest meetings, demonstrations, participation in *Panchayat* elections, and alliance building at state and national levels to force *Adivasi* rights onto the political agenda. These actions reinforce and create an active narrative of Adivasi political identity.

Despite these struggles, abrogation of rights is common. Fifty percent of claims under the FRA are rejected, and provisions in the Act that explicitly mandate the determination of rights prior to displacement are violated with impunity. However, many proposed mines are under legal challenge in 2019 or remain heavily contested and delayed by other means. In February 2019, the Federal Environment Ministry gave provisional approval for the Parsa mine in the Hasdeo-Arand, breaching the conditions that there would be no more mining approvals in the forest. A total of 150 *Gram Sabhas* in the area met to declare their opposition, planning a march to the state capital with the slogan of "keep your promises," *Vaada Nibhaao*, directed at the new Congress government. The villagers and activists seem to know that their actions, at most, will delay rather than stop mining operations in a context where illegal appropriation of land and despoliation of forests is increasingly common. However, as in the Lausitz, delay could help in the long-term, but there seems little sense that a renewable transition could relieve the pressure or change the politics.

As well as approving more coal mines and mine extensions, however, the Indian government is taking a leading role in promoting renewable energy. In Pavagada, in Karnataka, where our current research is taking place, 13,000 acres of land has been leased by many private players for roughly three decades from *Adivasi* and other villagers. Work on the solar park began in October 2016 with 600 MW of power commissioned by January 31, 2018. However, the villagers do not have copies of the lease documents, nor have the companies established the infrastructure (schools, colleges, and dispensaries) which had been promised as part of the agreement. Most importantly, there is no certainty that, at the end of the lease, the land now covered with a massive number of iron rods embedded in concrete to support the panels will be remediated or farmable. In this situation, there seems little room for public input into energy decision-making, design, or implementation. Thus far it would appear that neoliberal monetization and the drive for profits create very similar problems for local communities to those generated by the fossil fuel sector and works against their active participation in energy governance.

5.3 Australia

The Hunter Valley, defined by the large catchment of the Hunter River, forms part of the coal-rich Sydney and Gunnedah basins in New South Wales. Coal mining has been significant since the beginning of the NSW colony over 200 years ago. The Aboriginal people of the region, Wanaruah and Gomeroi, are not known to have used coal but there are Dreaming stories about the rock. Early mines, producing thermal coal that fed state-owned power generators, were underground and required a large labor force. They were the main employer in many towns and villages, coexisting with substantial agricultural land use, such as cropping, dairying, pastoralism, and more recently, wine growing and horse-breeding.

By the 1960s, coal, along with iron ore, became part of a burgeoning export industry supplying growing East Asian economies. This increase in demand led to large-scale open-cut mining and investment by transnational companies, which began to have notable impacts on other rural industries and on local lives. Nowadays, more than two-thirds of the upper Hunter Valley is under mining exploration leases and another 20% is occupied by mines. From the air, much of the area looks like a gigantic coal pit.

As coal became the dominant export, NSW state planning regulations became progressively more centralized. Conflicts about new and expanded coal projects have intensified, with rural producers and residents providing strong opposition to encroaching destruction of land. Coal policy continues to be expansionist, progressively taking power away from local people and councils in support of coal mining, despite an increasing threat to limited water supplies and local residents (Connor, 2016a,b). There are now more than 40 open-cut coal pits in the Hunter Valley, greatly reducing viable agricultural land use and impacting air quality, health, visual amenity, noise, light, biodiversity, and water supply, the last of pressing concern given the current severe drought (McManus and Connor, 2013). Wanaruah people's sacred sites and heritage are consumed by mining: "We are copping an absolute hammering from industry at the moment," one Traditional Owner exclaimed. Implementation of the conditions of consent are lax, and the NSW Environmental Protection Agency only weakly acts against breaches.

An integrated energy policy, which recognizes climate change, is still lacking at both State and Federal levels. The projected closure of the Liddell coal-fired power station in the lower Hunter Valley, when its owner AGL declared the plant uneconomical and that it was planning to transform the existing infrastructure into a renewable energy hub, lead to attempts, by the Federal government to persuade AGL to keep it open. The NSW Minister for Energy and Environment recently stated that legislation is being prepared to prolong the life of the state's coal-fired power stations and support new mines (2GB, 2019).

Conversations with local councilors in the Hunter and elsewhere reveal that confusion and regulatory difficulties inhibit, and sometimes stop, renewable development. In this environment, citizen involvement can be vigorous but can also seem sporadic and uncoordinated outside of a few organizing bodies, which often depend on a small core membership. Townsfolk (rural and urban) can easily be split between narratives of coal as a needed source of income and township stability and narratives of coal as destructive and poisonous. Protest against the destructive effects of coal is far more apparent than agitation for renewable energy.

Prominent examples of agitation against coal include the Anvil Hill coal mine protests in which locals allied with Greenpeace; the protests by horse breeders, winegrowers, and local residents against Drayton South mine extension, which used narratives of high status agriculture being destroyed unnecessarily; and the Rising Tide annual "blockade" of coal ships in Newcastle Harbour (the biggest coal port in the world). Legal representatives for "Groundswell Gloucester" managed to persuade the NSW Land and Environment Court that Gloucester Resources' proposed Rocky Hill mine would not only "cause significant planning, amenity, visual and social impacts" but that "the GHG emissions of the coal mine and its coal product will increase global total concentrations of GHGs" (quoting Judge Brian Preston). This is the first time an NSW court has recognized climate change as a reason to stop a mine (McGowan and Cox 2019). The company decided not to appeal (Hannam, 2019).

Citizens' organizations like "Hunter Renewal," the "Hunter Energy Transition Allianc," and CLEANaS (Clean Energy Association of Newcastle and Surrounds) work to mobilize people to plan for the postcoal future of the Hunter Valley. Hunter Renewal describes itself as a "project to bring people, businesses, and organisations of the Hunter Valley together to envision a diverse, resilient, and thriving future for our region." CLEANaS has attempted to introduce Green Bonds to the Hunter and has successfully funded a solar installation at Hunter Wetlands Centre.

Despite apparent hostility to transition from State and Federal Governments, local Councils have led the way in terms of action, with a meta-organization of councils, the "Cities Power Partnership," organized by the Climate Council.[1] For example, Newcastle Council, in the Hunter, has criticized the Federal Government's lack of support for renewable energy targets (ABC, 2015); voted to dump fossil fuel investments (Ryan, 2015); started to build a 5-MW solar farm at the Summerhill Waste Management Centre using a $6.5 million loan from the Clean Energy Finance Corporation; is replacing its cars with electric vehicles; and has announced it is going completely renewable from 2020. Lake Macquarie Council has likewise led renewable energy and local sustainability programs for more than a decade (Connor, 2016b).

While the Hunter Valley has few wind farms, there is a high rate of small and medium solar installation, but utility level electricity generation is almost exclusively coal based. Solar and wind generation makes up a very small percentage of total generation in NSW. Given the large area of land now devastated by mining in the Upper Hunter, large renewable projects in the areas may pose a further threat to agricultural land use, as in India, while coexistence of mining and solar farms may be difficult due to coal dust impairing panel efficiency. In other areas of NSW, strong opposition to solar farms has come from heritage tourism localities where the renewable landscape is considered a negative visual amenity. This could also

[1] The Climate Council originated when the Australian Climate Commission was abolished by the Government. It has since supported itself via crowd funding and donation, showing the popularity of climate action.

become a problem in the wine growing areas of the Hunter. In NSW's weak regulatory environment, neoliberal "market based" solutions dominate energy provision. While the industry-government nexus is weaker for renewables than for coal, the protection of rural environments and communities similarly seems overridden by the imperative of accumulation.

Despite hostility from governments and primarily neoliberal transition plans, renewables enjoy broad popular support, which rarely seem to translate into votes. The NSW Office of Environment and Heritage reported (2015: 49) that in the Hunter:

- Ninety-three percent supported using renewables to generate electricity in NSW.
- Eighty-five percent believed that NSW should increase the use of renewables over the next five years.

Some significant local narratives suggest less active support for renewable installation than an expectation that coal mining is dying and that the Hunter needs to be prepared for that transition. A survey of residents of the town of Muswellbrook found:

The only issue which received majority support (71%) was that a transition from coal would have significant effects on Upper Hunter Valley communities. While some positive impacts were mentioned, overwhelmingly concern was expressed about the economic effects of job losses and the flow-on effects of people moving from the area (Roden, 2018: 1)

Opposition to coal mining in the Hunter Valley can contribute to a wider narrative process of creating an organized politics against fossil fuel energy and, indirectly, for energy transition, but it has a long way to go.

5.4 Conclusion

In each of the case study locations where we conducted research, citizens, and local communities have actively participated in processes of energy governance, through the contestation of coal mining. In the Lausitz and in Chhattisgarh, they were able to stop some coal projects altogether or substantially delay their commencement. In the Hunter Valley, they were able to establish an important legal precedent explicitly linking coal extraction to climate change. In the Lausitz, coal opponents framed their resistance as a defense not just of their villages but as part of the *Energiewende*; energy transition functioned as an alternative narrative to continued coal extraction as a source of economic stability and regional identity. It is less clear that such an alternative narrative comes into play in the Australian and Indian cases.

When we turn to our case studies of actual energy transition, preliminary research suggests that they, too, reveal significant problems of participation and energy governance, which cannot be glossed over. While opposition to wind energy in Brandenburg is not organized on the same scale and does not have the same links to national and transnational environmental organizations as anti-coal activism, local people who feel they have not been given a stake in decision-making processes may actively oppose the expansion of wind energy, slowing down the pace of energy transition. In the Hunter region, while local councils have taken some important initiatives to promote a transition to renewable energy, these lack both support from state and federal government policy frameworks, and meaningful mechanisms for involving citizens in governance processes for energy transition. The latter is also true of Karnataka, although in India investment in renewables is actively supported by the central government but carried out by corporations in ways that appear to deliberately exclude local people from understanding, or participating equitably, in the process.

While Germany's Coal Commission has set a deadline of 2038 for an exit from coal extraction and coal-fired power, and India now has an Energy Transitions Commission, which describes itself as "a unique, high-level, multi-stakeholder platform on energy and electricity sector transitions in India" focused on "decarbonising the power sector" (ETI, 2018), Australia has no such blueprint and only loose coalitions of stakeholders advocating for transition. Moreover, in all three countries, the energy transition is occurring within the context of a neoliberal policy framework and is governed by neoliberal policy instruments such as Germany's system of reverse auctions for renewable energy or Australia's National Electricity Market. Such frameworks allow little opportunity for citizens' active participation in energy governance and may actively work to exclude them.

Over and above these considerations, our research thus far suggests a fundamental paradox: a more democratic approach to energy governance may not be fast enough to produce the necessary transformation of national energy systems in time, but a speedier transformation may be alienating for most people, put in place without proper consultation or participation, and generate protest and disruption.

These problems may well call for an "experimental politics" in which, instead of proposing a hard line policy in advance and sticking to it no matter what, we change policies according to the full range of results we observe, particularly attending to results which were unexpected or unintended, as this attention tells us more about how the system works and what needs to be done. Human views of reality are always partial, and we are dealing with the dynamic intersection of many complex and uncontrollable systems: ecological, climatic, economic, cultural, and so on. We have to give up illusions of total control and become aware of the limits of information, the resistance of established powers, and the presence of what appears like paradoxical, or contradictory, consequences of actions. Only then we are likely to have a relatively open and successful transition.

References

2GB, 2019. 'Coal is here to stay': State Government's major energy announcement. Matt Kean interviewed by Ray Hadley. <https://www.2gb.com/coal-is-here-to-stay-state-governments-major-energy-announcement/> (accessed 11.09.19.).

ABC, 2015. No compromise on RET puts jobs at risk: Nelmes. ABC News. <https://www.abc.net.au/news/2015-04-15/no-compromise-on-ret-puts-jobs-at-risk3a-nelmes/6393308> (accessed 15.04.15.).

Ahmad, N., 2014. Colonial legislation in postcolonial times. In: Lahiri-Dutt, K. (Ed.), The Coal Nation: Histories, Ecologies and Politics of Coal in India. Ashgate, Surrey & Burlington, pp. 258–260.

Climate Transparency, 2018. G20 Brown to Green Report 2018. <https://www.climate-transparency.org/g20-climate-performance/g20report2018>.

Connor, L., 2016a. Energy futures, state planning policies and coal mine contests in rural New South Wales. Energy Policy 99, 233–241.

Connor, L., 2016b. Climate Change and Anthropos: Planet, People and Places. Routledge, Abingdon.

Fabra, N., Matthes, F.C., Newbery, D., Colombier, M., Mathieu, M., Rüdinger, A., 2015. The Energy Transition in Europe: Initial Lessons from Germany, the UK and France. Towards a Low Carbon European Power Sector. Centre on Regulation in Europe (CERRE), Brussels. Macdonald.

Ghosh, D., 2018. Risky fieldwork: the problems of ethics in the field. Energy Res. Soc. Sci. 45, 348–354.

Government of India, 2016. Draft National Electricity Plan. Ministry of Power. Central Electricity Authority. Available from: http://www.cea.nic.in/reports/committee/nep/nep_dec.pdf.

Goodman, J., et al., 2020. Beyond the Coal Rush. A Turning Point for Global Energy and Climate Policy? Cambridge University Press, Cambridge.

Greenpeace India, 2012. How Coal Mining Is Trashing Tigerland. Greenpeace, Delhi.

Hannam, P., 2019. Key climate ruling against coal mine stands after miner declines to appeal. Sydney Morning Herald. <https://www.smh.com.au/environment/climate-change/key-climate-ruling-against-coal-mine-stands-after-miner-declines-to-appeal-20190508-p51ld1.html> (accessed 08.05.19.).

Heinrich Böll Stiftung, 2018. European structural change. First meeting of the trilateral group in Paris. <https://www.boell.de/en/2018/09/18/european-structural-change-first-meeting-trilateral-group-paris?dimension1 = division_oen>.

International Energy Agency, 2019. Perspectives for the Clean Energy Transition. IEA, Paris, <https://www.iea.org/publications/reports/PerspectivesfortheCleanEnergyTransition/>.

Kern, R., Rogge, S., 2017. The pace of governed energy transitions: agency, international dynamics and the global Paris agreement accelerating decarbonisation processes? Energy Res. Soc. Sci. 22, 13–17.

Lahiri-Dutt, K., Krishnan, R., Ahmad, N., 2012. Land acquisition and dispossession: private coal companies in Jharkhand. Econ. Polit. Wkly. 47 (6), 39–45.

MacDonald-Smith, A., 2018. Unstoppable force of clean energy transition not controlled by government policy. <https://www.afr.com/companies/energy/unstoppable-force-of-clean-energy-transition-not-controlled-by-government-policy-20181218-h1995f>.

McGowan, M., Cox, L., 2019. Court rules out hunter valley coalmine on climate change grounds. The Guardian. <https://www.theguardian.com/australia-news/2019/feb/08/court-rules-out-hunter-valley-coalmine-climate-change-rocky-hill> (accessed 08.02.19.).

McKay, H., 2019. BHP's economic and commodity outlook. <https://www.bhp.com/media-and-insights/prospects/2019/08/bhps-economic-and-commodity-outlook> (accessed 20.08.19.).

McManus, P., Connor, L., 2013. What's mine is mine(d): contests over marginalisation of rural life in the Upper Hunter, NSW. Rural Sociol. 22 (2), 166–183.

Ministry of Tribal Affairs, 2013. Statistical Profile of Scheduled Tribes in India. Government of India, New Delhi.

Morton, T., Müller, K., 2016. Lusatia and the coal conundrum. the lived experience of the German Energiewende. Energy Policy 99, 277–287.

Müller, K., 2019. Mining, time and protest: dealing with waiting in German coal mine planning. Extr. Ind. Soc. 6, 1–7.

Müller, K., Morton, T., 2018. At the German coalface: interdisciplinary collaboration between anthropology and journalism. Energy Res. Soc. Sci. 45, 134–143.

NSW Office of Environment and Heritage, 2015. Community attitudes to renewable energy in NSW. <https://www.environment.nsw.gov.au/resources/actionmatters/community-attitudes-renewable-energy-150419.pdf>.

Roden, J., 2018. Muswellbrook: between eight coal mines and two power stations. NSW Nurses and Midwives' Association. <http://www.nswnma.asn.au/wp-content/uploads/2018/07/Muswellbrook-between-8-coal-mines-and-2-power-stations.pdf>.

Ryan, P., 2015. Newcastle Council abandons fossil fuel investments. ABC AM. <https://www.abc.net.au/news/2015-08-27/newcastle-abandons-fossil-fuel-investments/6728900>.

Smith, A., 2019. BP says coal demand is stubbornly resistant in the face of surging renewables. Australian Financial Review, Feb 11 2019, https://www.afr.com/companies/mining/bp-says-coal-demand-is-stubbornly-resistant-in-the-face-of-surging-renewables-20190215-h1baie. Energy Transitions Commission India. https://www.teriin.org/energy-transitions.

UN News Centre, 2019. Climate action: 4 shifts the UN Chief encourages governments to make. <https://news.un.org/en/story/2019/05/1038381>.

CHAPTER

6

Cosmopolitan governance for sustainable global energy transformation: democratic, participatory-deliberative, and multilayered

Andreas Klinke

Environmental Policy Institute, Memorial University of Newfoundland, Canada

OUTLINE

6.1 Challenges of global energy transformation

A majority of scholars and experts worldwide believe that a transformation of current energy systems to a more sustainable practice is necessary if the risks of climate change are to be reduced or mitigated (Intergovernmental Panel on Climate Change (IPCC), 2014). However, the Organization for Economic Co-operation and Development (OECD, 2012) predicts that there will be no perceptible change in the global energy mix of the future, with fossil energy remaining at about 85% of the total energy used, without new energy

The Role of Public Participation in Energy Transitions
DOI: https://doi.org/10.1016/B978-0-12-819515-4.00006-4

policies and governance structures. For this reason, societies worldwide are increasingly confronted with the task of coping with profound energy transformation away from the use of fossil energy demanded by traditional modern production implemented since the Industrial Revolution. Instead new guiding principles and worldviews that entail renewable energy generation and consumption, low-carbon economies, and sustainable global climate policies will have to shape the future. Such an energy transformation would spark major challenges to traditional structures in all political, social, and economic spheres and would bring about social upheaval and structural change.

The scope of such an energy transformation would have a profound effect on established socio-technical systems that foster high-carbon economies, whether technologies, institutions, economic sectors, culture, or science (Wissenschaftlicher Beirat der Bundesregierung Globale Umweltveränderungen (WBGU), 2011: 93). The global challenge is to gradually reverse the overuse and exploitation of natural resources that is fueled by an undampened greed for fossil energy resources. The drive for growth in a globalized and unrestrained market economy will need to give way to a more complex, sustainable development of the earth system and global energy production. A crucial component of energy transformation is, indubitably, the transfiguration of culturally embedded orders of energy use encompassing a range of issues that are conveyed through social and political learning. As a result, challenges that manifest such a transformation feature profoundness and scope of cross-border nature that a national framework of rules and directives cannot address the issues on its own and gain control of; the national capacity for action is overburden. In other words, global energy transformation causes global uncertainty and challenges the fundaments of socioeconomic organization, security, ethics, and worldviews. For example, renewable energy production grew over 40% in Germany in 2018 for the first time; it increased almost 8% against the previous year (Frauenhofer Institute for Solar Energy Systems (ISE), 2018). Although each country ought to navigate its own energy transition, no single nation-state is capable of tackling global energy issues on its own because domestic politics and regulations are too insignificant to achieve political outcomes that can guide a global energy transformation. Hence some scholars call for an energy transformation governance that is capable of steering a sustainable large-scale transition away from fossil fuels, beyond the scope of any single national government (cf. Scrase and MacKerron, 2009; Florini and Dubash, 2011; Meadowcroft, 2011; WBGU, 2011; Strunz et al., 2015). Yet our world is driven by the domestic regulations of nation-states, whereas a global energy transformation requires postnational, multilayered global energy policy, politics, and polity from the local level to the global level.

Sustainable global energy transformation to mitigate global climate change impacts is a Herculean task that produces a coercive inclusion of the global multitude which is what I understand as the cosmopolitization of energy transformation (cf. Beck, 2006, 2013; Beck and Sznaider, 2006). It concerns the entirety of national societies, the global community, and world politics and comprises interconnected interactions, processes, and structures. The cosmopolitization of energy transformation disengages from the centrality of nation-states as central actors in international politics. It facilitates new actor and power constellations; state, civil society, and economic actors become more equal partners that expedite sustainable global transformation as a cosmopolitan entity. I argue that essential qualities that are democratic, participatory-deliberative, and multilayered, performed through a functional division of "specialized authorities" (Burnheim, 2006: 17) stand for cosmopolitan governance as a way of navigating and mastering a sustainable global energy transformation that is of a normatively and ethically higher quality than other forms of governance. Hence my proposal for cosmopolitan governance for sustainable global energy transformation is based on three specialized sites of authority: expert and scientific knowledge, group-based practical experience and intelligence, and public wisdom and vision through transnational public spheres. My proposal encompasses these three forms of discourse and the participation of and deliberation between experts, scientists, collective actor groups, and the world at large. Distinctive, yet intertwined and complementary, the three authorities help solidify transnational communication and decision-making on energy transformation.

6.2 Essential qualities of cosmopolitan governance

Great transformations such as a sustainable global energy transformation would be best guided by democratic culture and institutions, but ironically, such a profound transformation also challenges traditional democratic practices. This chapter does not offer an analysis of whether and to what extent the current energy transformation in Germany or worldwide is democratic. Hence I use the attribute "democratic" in this chapter as a way to communicate making global governance more democratic toward cosmopolitan governance for sustainable global energy transformation. I probe how political power can be democratically legitimated beyond the nation-state. How can international decision-making on global energy transformation be justified, and who should be entitled to participate in the shaping of such global transformation? I discuss what kind of democratic institutions and processes are most reasonable for guiding sustainable global energy transformation. My attention is focused on which approach to cosmopolitan governance produces adequate democratic authority to master and navigate sustainable energy transformation and what is reasonable to expect from citizens in transition periods. I also approach the question of why one democratic form is more morally desirable than others. As we will see, classical representative democracy is not compatible enough with what we expect from citizens and collective actor groups in terms of providing concrete guidance for a more renewable energy production and consumption.

The democratic nature of cosmopolitan governance is not associated with representative, equal, and unweighted voting. Rather it is in reference to globally democratic institutions and processes of scientific justification and public reasoning. The idea behind this approach is that international policies and regulations on sustainable global energy transformation are legitimate when they have been publicly debated and justified by experts and the global multitude across borders. Many scholars and advocates of modern democracy think that participation and deliberation involving individuals and groups is an important avenue for ordinary citizens and group representatives to exchange perspectives with others through argumentation. Public, democratic deliberation gains traction and is respected when respectful face-to-face communication that entails a diversity of viewpoints takes place. A plurality of methods and critical approaches as well as the pooling of facts and information in public deliberation are essential for making cosmopolitan governance democratic and to successfully solve problems for the common good (cf. Freeman, 2000; Klinke, 2009; Landemore,

2011; Mansbridge et al., 2012). One example of such participation and deliberation is the democratized transnational environmental governance of the Great Lakes by Canada and the United States that has been ongoing for many years (Klinke, 2006, 2009, 2012).

Public participation and deliberation in cosmopolitan governance means either the direct involvement or representation of the public and collective actor groups (such as nongovernmental organizations, business organizations, unions, etc.) in international decision-making and international public policies that concern global energy transformation. By participating in public deliberations, people without societal political rank and collective actor groups engage in careful consideration together with international experts, government agencies, political leaders, and international organizations to shape international public policies and programs pertaining to global energy issues (cf. Öberg, 2016; Quick and Bryson, 2016).

Drawing on Parkinson and Mansbridge's edited volume "Deliberative Democracy at the Large Scale" (2012), I argue that participation and deliberation in cosmopolitan governance couples classic democratic ideals with contemporary pragmatic elements of expanded deliberation. Participation and deliberation advances public discourse and communication, allowing actors to learn about the preferences and values of others. Participation and deliberation should be open to all people affected by energy transformation and "participants should have equal opportunity to influence the process, have equal resources, and be protected by basic rights" (Mansbridge et al., 2010: 65). Participation and deliberation induces actors to listen to each other and justify their interests and opinions in a way that they believe is good and right and then attempt to arrive at a reasoned consensus on the best option based on what is socially and publicly acceptable. Such communication focuses on mutual justification, that is, participants accept considerations that are justified and persuasive, even if they disagree with them. Since the plurality of values and interpretations expressed in participation and deliberation brings out self-interest and conflicts of interest, the process allows constrained self-interests, conflicts of interest, and noncoercive forms of deliberative negotiation (Mansbridge et al., 2010). The outcomes of such a process include various forms of communicative agreement such as reasoned convergence and consensus, incomplete agreement, negotiated compromise, and agreed account for a dissent. Participation and deliberation are the democratic avenue in cosmopolitan governance that would enable a functional division of specialized authority at multiple layers, as we will see.

Democratic forms of governance emphasize the multilevel or multilayered or multitiered structure and organization of a polity. Such forms labeled as multilevel, multilayered, or multitiered reflect a diversity of interactions, processes, and institutions that are increasingly difficult to grasp clearly. Cosmopolitan governance as multilayered governance, as I denote it from here on, is a combination of different forms of governance that represent complex webs of vertical and horizontal relations. The attribute "multilayered," describing one of the major qualities of cosmopolitan governance for sustainable energy transformation, refers to both the classical understanding and newer interpretations of multilevel governance (cf. Hooghe and Marks, 2010; Bache et al., 2016) that differently underline the significance of the vertical or horizontal dimensions. To further differentiate the vertical from the horizontal arrangement of multilevel governance, Hooghe and Marks

(2003, 2010) elucidate a bipartite typology that brings together previously discrete theories related to governance.

The first type is the sharing and hierarchical coordination of governing power between national and subnational governments in a federalist structure. This classical, federalist approach is inadequate as a means to handle the challenges of transformation because it remains the shadow of classical power and hierarchy that has failed repeatedly to steer transformations. For example, European energy transformation with its multilevel governance approach exhibits the tension between the centralized idea of a common European market integration and the member states' autonomy to develop policies according to domestically legitimized interests (Tews, 2015). The difficulty of coordinating energy transformation and climate policy in the EU is due to a complex federal system with given traditional power hierarchies that enables policy making in a pluralistic way and at multiple political levels but does not go beyond the lowest common denominator of overlapping national interests. The European multilevel governance system fails to take into account that the judicial and political bodies of municipalities, constituent states, and nation-states are in a linear relationship with each other. Some decisions on energy transformation and policy are made at the European level, others at the national level, and some at the subnational and regional levels. Yet the governing of European energy transformation is erratic, more a patchwork than a collective transition process aiming for common interests. National paths diverge considerably in terms of goals, pace, and scope; the transformation is unintegrated and not consistent. This kind of federal multilevel system undermines the traditional forms of representation but does not produce new mechanisms.

I argue that cosmopolitan governance for sustainable global energy transformation aligns with the second type of multilevel governance outlined by Hooghe and Marks (2003, 2010), which emphasizes the reallocation of political authority to a nonhierarchical and polycentric structure involving a wide range of public and private actors (cf. Zürn, 2010; Bache et al., 2016). Hence cosmopolitan governance delineates a multilayered configuration that evolves beyond subnational and national boundaries but not in the form of the federalist governance. A newer interpretation of dynamic multilevel governance that has inspired and illuminated the multilayered character of cosmopolitan governance for sustainable global energy transformation is a combination of different forms of governance that represent complex webs of vertical and horizontal relations and interactions (Klinke, 2017). It highlights the increasing vertical interactions and processes among different territorial and jurisdictional levels *and* the intensified horizontal interactions between state, civil society, and economic actors in a democratic network as a means to navigate complex challenges arising in the course of energy transitions. Energy transformation requires navigating new configurations and responsibilities in such a way that the authority to make decisions is transition specific and has a transformative and structuring capacity to influence the course of transitioning in the vertical and horizontal dimensions. Indeed this authority operates in overlapping levels. It is a kind of unbundling of centralized state power into a broader spectrum of vertically and horizontally organized entities that emphasize dynamicity and flexibility. Multilayered governance draws the focus to cosmopolitan governance as a network without disregarding the different levels of government; it attempts to capture the dynamics of state–society relationships and the interplay of functional and territorial governance. Furthermore it is important that multilayered governance is flexible and adaptable in terms of vertical and horizontal scales and that it captures the consequences of corresponding energy policies. Decentralized authority and responsibility can better reflect the diversity of interests and preferences of ordinary people and collective actor groups than centralized ones.

6.3 Knowledge-based authority

What exactly do we know, and what not, about global energy transformation that is more ecologically, economically, and socially sustainable? What do we need to know about sustainable global energy transformation and its governance? Knowledge about a sustainable global energy transformation is one way of getting at a paradigm that offers new guiding rational beliefs and a worldview that advances a sustainable energy supply on a global scale. Therefore the production of knowledge concerns the articulation of what exactly the challenges are of sustainable global energy transformation and how to address them adequately.

Addressing energy-specific topics, both domestic and international, requires experts with comprehensive and authoritative knowledge and skills, whether that is in regard to renewable energy technology, infrastructure matters, and trade-offs between centralization and decentralization. This specialized knowledge is necessary to shift beliefs about energy transition. Indeed the opinion and will formation of the public cannot begin without there being an assessment about a correct or good path toward more sustainable energy systems. Evidence-based statements, reliable scientific studies, and the most recent specialized state of knowledge are important to this end. In these cases, laypersons trust experts whose opinions are plausible and reasonable. Expert advisory bodies, institutes of higher education, independent research institutes, nonprofit organizations, impartial think tanks, and independent state-run research agencies provide the relevant issue-specific expertise and competence. By connecting in epistemic communities and institutions, they are able to collectively facilitate agreement on cognitive and normative ideas about global policy-relevant problem-solving (cf. Haas, 1992, 2011, 2014). Global epistemic communities and institutions are needed for sustainable global energy transformation because only they can acquire the specialized knowledge and thus substantive authority needed to address transnational and international issue- and task-specific challenges. Only they can generate reliable and profound scientific knowledge that is policy-relevant for interdependence, global socioeconomic change, and globalization of energy-related issues, especially when it comes to technical knowledge.

Hence, cosmopolitan governance for sustainable global energy transformation acquires knowledge-based authority that involves scientists and broad epistemic communities shaped by collaboration and competition. Scientific experts from pertinent scientific institutions hold the specialized knowledge and skills needed to address the cognitive and evaluative problems that arise from the scientific challenges of a sustainable global energy transformation. They are able to establish the most cogent explanation and clarify dissenting views with regard to causal beliefs. They facilitate deliberation among experts of formally acknowledged research institutions and are able to advance the state of the art in respected knowledge domains. It is important that they are recognized as representatives of their respective scientific communities, able to provide professional expertise that generates cognitive and evaluative knowledge relevant for reference frames and meaning structures in terms of global sustainable energy transformation. They are also able to validate ideas about and criteria for the evaluation of desired transformation. Research units possess substantial authority because they operate with a sense of credible obligation when it comes to the objective and unprejudiced production of expert knowledge and systematic information that is generally accepted in the public sphere. The overall goal is to establish consensual knowledge about cause-and-effect relationships, uncertainties and ambiguities, and policy-relevant criteria for judging societal acceptability and tolerability. Since truth seeking motivates scientific experts, communication and deliberation aim at cognitive convergence.

Drawing on such knowledge-based authority would allow cosmopolitan governance to investigate the processes and structures of a global transition from traditional modern production, with its use of fossil energy, to renewable energy generation and consumption and the development of low-carbon economies. It would seek to outline how regional, national, and global advancements could proceed, impediments notwithstanding. Such a knowledge-based authority would need to gather data and elicit country-specific reporting on a variety of domains, including new political initiatives and legislation, emerging technologies, infrastructural change, extraction of fossil resources, energy trading, and so on. It would also need to work hand in hand with the IPCC on its work on sustainable global climate policy.

International, institutionalized knowledge-based authority is essential if cosmopolitan governance is to develop new ways of producing global sustainable energy and justify why they better advance a sustainable future of humankind on earth. Such a paradigm change would come about with the clarification of causal and principled beliefs and worldviews (cf. Goldstein and Keohane, 1993). These foundations would help the public understand the new underlying reference frames and meaning structures. Causal beliefs provide evidence of causal inferences, for example, that there is a connection between a renewable energy supply and a decrease in carbon emissions. Principled beliefs are expressed in ideational norms and rules that help us distinguish between right and wrong and just and unjust when, for example, evaluating how much fossil energy is acceptable to use in terms of climate change reduction targets and what is a reasonable distribution of renewable and fossil energy in a transition period. It is also important to make the public aware of conflicts and moral disagreements that arise when traditional perspectives are at odds with progressive and innovative action. Additionally we learn from a knowledge-based authority that principled beliefs, even conflicting ones, are embedded in larger belief systems or worldviews. Worldviews in this regard are based on value systems that include views about the economic and social organization of human existence. The beliefs that make up worldviews are not detached from each other but rather interdependent: "Causal beliefs imply strategies for the attainment of goals, themselves valued because of shared principled beliefs, and understandable only within the context of broader world views" (Goldstein and Keohane, 1993: 10).

Ensuring that global energy transformation is based on international knowledge-based authority is essential for three reasons. First it allows for tools to develop an integrated global perspective on current and future energy supplies that is built on common sense and scientific images of the worldwide situation. Second it can serve as a scientific platform on which to build commonly accepted transformative knowledge and expertise on global energy and facilitate interdisciplinary research on sustainable energy that is not dominated by the traditional paradigm. Third it facilitates interdisciplinary collaboration on renewable technology development and infrastructure solutions and enables the cultural and ethical imagination needed to shape the extensive socioeconomic changes engendered by transformation.

New cosmopolitan governance for sustainable global energy transition necessitates knowledge-based authority, structured in part like the IPCC. Although the epistemic status of the IPCC is disputed, I use it as an example to illustrate how an expert deliberation system has been established as a focal point of a global epistemological discourse. The IPCC and its epistemic community has addressed the multifaceted complexity of climate change and facilitated a reasoned and internationally accepted consensus of the causality between human behavior, for example, fossil energy production and consumption, and increased global warming, even though considerable scientific uncertainty remains about the consequences. The IPCC's mission and goal is to provide cognitive and evaluative frames of reference and meaning that help inform and guide public and political opinion and decision-making.

Cosmopolitan governance with a knowledge-based authority would be arranged as an independent, international system of science and expert advice—a scientific community—not as an intergovernmental panel guided by national governments. It could consist of distinct working groups or task forces. One working group could evaluate countries' progress along the path of renewable energy and combine the varying status quos in countries into a coherent whole as a way to review the state of global energy transformation. Another group could estimate the adaptability of existing infrastructure and socio-technical/economic systems to energy transformation. A third working group could appraise economic, social, and political options for mastering the challenges of energy transformation through structural and social-economic change. The cognitive and evaluative reference frames and meaning structures produced in the matrix of knowledge-based authority enter the participatory-deliberative process of collective actor groups and transnational public spheres as valid and reliable scientific substance.

6.4 Experiential authority

Experiential authority grows out of practical knowledge and skills as well as real-world know-how over a period of time. In the context of global energy transformation, experiential authority is established when groups of people organize for a joint purpose to rationally consider fundamental issues and ethical challenges of sustainable energy production and consumption—the paths to achieving a more sustainable being and way of life, the necessary technical and infrastructural systems to bring about new ways of being, and the kinds of public services required. Such deliberation creates a space for reflexive communication in which members of relevant collective actor groups, such as nongovernmental organizations, businesses, utility companies, energy providers, unions, and political authorities, exchange ideas and lessons learned from their world. Thus these groups become agents of change who critically scrutinize the limits of established paradigms and produce a narrative and principles of a renewable and sustainable global energy that could be commonly accepted and trusted. Drawing on knowledge-based authority, the collective actor groups deliberate on their experiences and interpretations of social and professional life and discuss their commonalities and differences on issues of sustainable global energy transformation. The goal of group-based participation and deliberation is to evaluate actions and behaviors by determining the acceptability and tolerability of economic and social options and risks that emerge in the course of the transfiguration of global energy orders.

Even though conflicts of interest and competing norms and values emerge, it is important that the participation and deliberation of these groups focuses on mutual justification and acceptance of conflicting experiences as well as plausible considerations that are compelling and persuasive even for those who disagree with them. It "opens the door to storytelling and the non-cognitive evocation of meanings and symbols that can appeal to actual or imagined shared experiences" (Mansbridge et al., 2012: 67), which is essential if moral questions arise justified by cultural values and reflective of ontological and ethical convictions. Narrative communication can establish credibility and mutual respect among the participants, which allows them to attend to both commonalities and differences. Thus the amalgamation of scientific and experiential substance creates an epistemic and moral surplus that enables the evaluation of global energy transformation on the large scale of a transnationalized public community of peoples, which is essential in the creation of cosmopolitan governance on sustainable global energy transformation.

The Ethics Council on Energy Transition established by the German federal government in 2012 in the aftermath of the Fukushima accident serves as an instructive example of how transnational collective actor groups through participation and deliberation can become legitimized experiential authorities paving the way for international decision-making on sustainable global energy transformation. The Council represents an intermediary and moral authority in a larger context of governance that facilitates communication between the state and society in a pluralistic and corporate manner, though its scope is limited to domestic politics. The Council comprises representatives from the scientific community, government, the economic sector, and civil society. It is associative and independent in nature. Council members appraised the risks and benefits of phasing out nuclear energy and in this process came to an agreement on a commonly reasoned policy that recommended a transformation to more renewable forms of energy. This recommendation has been unanimously approved by the German parliament.

6.5 Transnational public spheres

The formation of public opinion about global energy transformation would be the most important vehicle of representation in the configuration of cosmopolitan governance. Public opinion is a form of mediation between numerous individual- and group-specific articulations of interests and claims, national societies, and an emerging world society. If individuals around the globe—a global multitude—participate in trans/internationalized communication about social, economic, political, and cultural issues of common concern with regard to global energy transformation, then trans/internationalized public spheres, appropriate entities for democratic agency across national boundaries, will emerge (Steffek, 2010; Klinke, 2014). These spheres unfold through the transnationalization of discourses and the transnationalization of "space for the communicative generation of public opinion" (Fraser, 2007: 7) that represents the mediative authority between people across borders and international political decision makers. Such a trans/internationalized communication relates to common contextual frames of reference in terms of observation, perception, action, and interdependence. Commenting on, opining on, or reacting to what other speakers say in a trans/international context shapes discursive arenas with a communicative formation of public opinion and will. Transnational communicative spaces have been identified within the EU, and, in particular, a transnationalization of public spheres has emerged in continental Western and Southern Europe (Risse, 2010). These transnational public spheres emerge "whenever European issues are debated as questions of common concern" (Risse, 2010: 6), such as European energy transformation.

Actors' engagement in trans/internationalized arenas of public opinion can lead to the establishment of a trans/internationalized public community of continuing discourse. If such discourses across borders identify international collective effort as the best means for handling sustainable global energy transformation, then the keystone for the creation and institutionalization of a cosmopolitan governance architecture has been established. Thus the trans/internationalization of public discourses segues into a globalized public community—a trans/internationalized democratic demos defined by its base of peoples across hierarchies and national boundaries without predestined jurisdictions—playing a more powerful and active role in a transnational, democratic polity (Klinke, 2014). A globalized public community becomes the primary avenue for the production of the transformative and structuring power needed to advance global energy transformation. This network of public deliberation legitimizes and provides the framework for cosmopolitan governance by developing common ground among peoples, defining ways of shaping a new paradigm of sustainable global energy supply, and creating opportunities to act in the new order.

The asymmetry of public opinion in the trans/internationalized public community would be countervailed through something like "*equal opportunity of access to political influence*" (Knight and Johnson, 1997: 280, italics in original) by adequate means of public participation and deliberation. Opportunities of public participation and deliberation within a cosmopolitan governance network system would become the locations of the production of transformative and structuring power. These participatory-deliberative processes could claim to be representative of the trans/internationalized public community to some extent but would not draw on statistical representativeness or electoral or proportional representation (cf. Goodin and Dryzek, 2006). However, the reflexive network of deliberation would ensure that social diversity, plurality of interpretations, and the underlying beliefs and values of the trans/internationalized public community were represented through direct involvement of those who are affected. This descriptive representation would serve as a vehicle for democratic representation (cf. Parkinson, 2006: 154–155). Hence cosmopolitan governance would supplant traditional territorial representation but not the democratic principle of majority rule.

Transnational public participation and deliberation could be organized by means of mini-publics in the form of consensus conferences, citizen juries or panels, and deliberative opinion polls comprised of ordinary citizens (cf. Goodin and Dryzek, 2006; Goodin, 2008; National Research Council, 2008; Fishkin, 2009; MacKenzie and Warren, 2012). Such forms of public participation and deliberation would establish a space where the forming of global public opinion and will could be channeled and aggregated, thus creating direct democratic practice extending beyond national hierarchies. Nonorganized affected individuals, selected by lot (sortation), would be entitled to state their experience and desires; they would also be authorized to co-influence international political decision-making. Participants in mini-publics reason together about decisions by exchanging narratives about and making claims on shared topics (cf. Parkinson, 2012: 154). Thus the global multitude would be the third specialized authority in the functional division of the cosmopolitan governance configuration. The goal of public participation and deliberation is that individuals have an equal opportunity to exchange their subjective perspectives and collectively discuss practical questions with regard to the steering and control of energy transformation. "The best discussions clarify both conflict and commonality, and perhaps forge genuine commonality where it had not existed before" (Mansbridge, 2006: 118). The participants legitimize standards for the handling of transformation through the claim that the new norms and rules deserve recognition because they are right and good and regulate the behavior of all in the common interest. Previous experiences with mini-publics reveal that ordinary citizens develop well-considered valuations that can solidify public opinion at a large scale, complement expert judgments, and formulate politically relevant policy (MacKenzie and Warren, 2012: 95). Hence public participation and deliberation would determine strategies to instruct specific international public policies addressing energy transformation. The specialized authority of mini-publics would strengthen the cosmopolitan governance configuration by creating trust relationships between transnational public spheres and executive agencies of governments and international organizations (cf. MacKenzie and Warren, 2012: 96–97). The global multitude would trust the mini-publics because they would serve as faithful custodians of the information and experience that guides peoples' political judgments. Governments and international organizations might trust mini-publics to help guide trans/international energy decision-making because they can anticipate public opinion on issues that have not yet been scrutinized, especially with regard to contentious issues arising from uncertainties and sociopolitical ambiguities in the context of energy transition.

6.6 Conclusion

Cosmopolitan governance for sustainable global energy transformation is charged with effectively dealing with trans/international challenges arising in the course of energy transitions that national institutions cannot tackle independently. My proposed model of cosmopolitan governance relies on a functional division of specialized authority—knowledge-based authority, experiential authority, and transnational public spheres—that is conveyed by democratic participation and deliberation and multilayered institutions. It goes beyond the limits of traditionally detached realms of domestic hierarchical politics and the intergovernmental regime and organization approach characteristic of world politics.

However, it neither implies a romantic vision of a political utopia nor a unified world government that holds sway over all humankind under one political authority. Rather it is an alternative, pragmatic approach paving the way to a cosmopolitan, democratic governance that is reasonable and realistic in that it references practical domestic and transnational considerations. One could argue that the notion of cosmopolitan governance is a genuine reflection of the collective necessity of a united humankind in the light of catastrophic anthropogenic climate change and the need for sustainable global energy transformation. It represents an innovation in global institutional design to move humanity toward more democratic world politics. It is an approach that offers an effective global transition toward more inclusion (especially of collective, nonstate actor groups and individuals), democratic self-determination, free forming of public opinion and political will, global justice, and freedom. These principles should guide international action in the context of the challenges of climate change and global energy transformation. In cosmopolitan governance, all people are equal members of a global community in one world where every human being, irregardless of national affiliation, has equal moral worth and democratic right and is entitled to an equal consideration of their interests. It emphasizes our global moral horizon and the global responsibility we have in light of anthropogenic climate change, even while national and local transformation of energy systems is needed. While conceptions of good life and prosperity are legitimate considerations in deliberating energy transitions, we can no longer overlook that we are all in one connected world. Cosmopolitan governance for sustainable global energy transformations draws the global responsibility closer to the local one and vice versa and aims to situate humans as members of overlapping spheres of responsibility in the matter of climate change and energy supply.

Participation and deliberation of collective actor groups in transnational public spheres in the form of mini-publics would be organized locally and interregionally. These processes would exist as permanent institutional mechanisms to channel and aggregate public opinion and will in an emerging transnational community of communication and discourse. They would share contextual frameworks concerning the scope of renewable energy systems. Thus they would be representative of direct democratic mechanisms of democratic aggregation. They would be polyarchic, group involvement of organized collective interest groups and a micro-cosmopolitan public participation and deliberation of laypersons and nonpartisan, unorganized individuals of transnational public spheres.

When human beings through participation and deliberation design global rules and practices for energy transitioning, "they ought to disregard their private and local, including national, commitments and loyalties to give equal consideration to the needs and interests of every human being on this planet" (Pogge, 2013: 298). My proposed cosmopolitan governance for sustainable global energy transformation is not in tension with domestic duties and commitments. The democratic principles of cosmopolitan governance would ensure members of the global community are treated as equals in their entitlements and allow a legitimate role for domestic energy policy and politics that operate within national constraints.

It can be argued that cosmopolitan governance makes the global multitude sensitive to a more cosmopolitan identification that we are all citizens of one world conjoined on a fundamental level to a global energy community that is in need of sustainable energy transformation. Hence this cosmopolitan governance serves to bring about more global renewable energy systems by enabling global processes and institutions that contribute to a sustainable common good and a worldwide energy transition. It suggests a point of unification for humankind to strive for a sustainable energy future. Cosmopolitan governance for sustainable global energy transformation does not derogate the traditional state power to alter domestic energy systems. "Rather it seeks to entrench and develop political institutions at regional and global levels as a necessary complement to those at the level of the state" (Held, 2003: 478). This occurs by reproducing democratically effective intrastate processes and institutions at the transnational and global levels, such as expert panels, group-based participation and deliberation, and public participation and deliberation by means of mini-publics. My model of cosmopolitan governance seeks the participatory and "*deliberative democratization*" (Klinke, 2016: 98, italics in original) of processes and institutions to help navigate sustainable global energy transformation from the local to the global. Cosmopolitan governance has an institutional and processual design that disperses sovereignty vertically and horizontally instead of concentrating political authority in all-encompassing international institutions. In this regard, we establish "a system in which the political allegiance and loyalties of persons are widely dispersed over a number of political units of various sizes, without any one unit being dominant and thus occupying the traditional role of the state" (Kleingeld and Brown, 2013). In my view, global political loyalty is not in tension or competition with national loyalty. The increasing public awareness of transnational and global issues that arise from anthropogenic climate change and traditional fossil energy systems is an ongoing process that includes both the development of political support, involvement, and commitment to national sustainable energy transitions and a wider sustainable global energy transformation.

References

Bache, I., Bartle, I., Flinders, M., 2016. Multi-level governance. In: Ansell, C., Torfing, J. (Eds.), Handbook on Theories of Governance. Edward Elgar, Cheltenham, pp. 486–498.

Beck, U., 2006. *The Cosmopolitan Vision.* Polity Press, Cambridge.

Beck, U., 2013. Methodological cosmopolitanism—in the laboratory of climate change. Soziologie 42 (3), 278–289.

Beck, U., Sznaider, N., 2006. Unpacking cosmopolitanism for the social sciences: a research agenda. Br. J. Sociol. 57 (1), 1–23.

Burnheim, J., 2006. *Is Democracy Possible? The Alternative to Electoral Democracy.* Sydney: Sydney University Press.

Fishkin, J.S., 2009. *When the People Speak: Deliberative Democracy and Public Consultation.* Oxford University Press, Oxford.

Florini, A., Dubash, N.K., 2011. Introduction to special issue: governing energy in a fragmented world. Glob. Policy 2, 1–5.

Fraser, N., 2007. Transnationalizing the public sphere: on the legitimacy and efficacy of public opinion in a post-Westphalian world. Theory Cult. Soc. 24 (4), 7–30.

Frauenhofer Institute for Solar Energy Systems (ISE), 2018. *Stromerzeugung in Deutschland im ersten Halbjahr.* https://www.ise.fraunhofer.de/content/dam/ise/de/documents/publications/studies/daten-zu-erneuerbaren-energien/ISE_Stromerzeugung_2018_Halbjahr.pdf.

Freeman, S., 2000. Deliberative democracy: a sympathetic comment. Philos. Public Aff. 29 (4), 371–418.

Goodin, R.E., 2008. *Innovating Democracy: Democratic Theory and Practice after the Deliberative Turn*. Oxford University Press, Oxford.

Goodin, R.E., Dryzek, J.S., 2006. Deliberative impacts: the macro-political uptake of mini-publics. Polit. Soc. 34 (2), 219–244.

Goldstein, J., Keohane, R.O., 1993. Ideas and foreign policy: an analytical framework. In: Goldstein, J., Keohane, R.O. (Eds.), Ideas and Foreign Policy. Beliefs, Institutions, and Political Change. Cornell University Press, Ithaca, NY, pp. 3–30.

Haas, P.M., 1992. Introduction: epistemic communities and international policy coordination. Int. Organ. 46 (1), 1–35.

Haas, P.M., 2011. Epistemic communities. In: Badie, B., Berg-Schlosser, D., Morlino, L. (Eds.), International Encyclopedia of Political Science. Sage Publications, London, pp. 788–791.

Haas, P.M., 2014. Ideas, experts and governance. In: Ambrus, M., Arts, K., Hey, E., Raulus, H. (Eds.), The Role of "Experts" in International and European Decision-Making Processes. Cambridge University Press, Cambridge, pp. 19–43.

Held, D., 2003. Cosmopolitanism: globalization tamed? Rev. Int. Stud. 29 (4), 465–480.

Hooghe, L., Marks, G., 2003. Unraveling the central state, but how? Types of multi-level governance. Am. Polit. Sci. Rev. 97 (2), 233–243.

Hooghe, L., Marks, G., 2010. Types of multi-level governance. In: Enderlein, H., Wälti, S., Zürn, M. (Eds.), Handbook on Multi-level Governance. Edward Elgar, Cheltenham, pp. 17–31.

Intergovernmental Panel on Climate Change (IPCC), 2014. *Fifth Assessment Report*. IPCC, Geneva.

Kleingeld, P., Brown, E., Cosmopolitanism, 2013. In *Stanford Encyclopedia of Philosophy*. https://plato.stanford.edu/entries/cosmopolitanism/#2.

Klinke, A., 2006. *Demokratisches Regieren jenseits des Staates. Deliberative Politik im nordamerikanischen Große Seen-Regime*. Opladen: Barbara Budrich Publisher.

Klinke, A., 2009. Deliberative Politik in transnationalen Räumen—demokratische Legitimation und Effektivität der grenzüberschreitenden Wasser- und Umweltpolitik zwischen Kanada und USA. Politische Vierteljahresschr. 50 (4), 774–803.

Klinke, A., 2012. Democratizing regional environmental governance: public deliberation and participation in transboundary eco-regions. Glob. Environ. Polit. 12 (3), 79–99.

Klinke, A., 2014. Postnational discourse, deliberation and participation toward global risk governance. Rev. Int. Stud. 40 (2), 247–275.

Klinke, A., 2016. Democratic theory. In: Ansell, C., Torfing, J. (Eds.), Handbook on Theories of Governance. Edward Elgar, Cheltenham, pp. 86–100.

Klinke, A., 2017. Dynamic multilevel governance for sustainable transformation as postnational configuration. Innovation Eur. J. Soc. Sci. Res. 30 (3), 323–343.

Knight, J., Johnson, J., 1997. What sort of equality does deliberative democracy require? In: Bohman, J., Rehg, W. (Eds.), Deliberative Democracy. Essays on Reason and Politics. MIT Press, Cambridge, MA, pp. 279–319.

Landemore, H., 2011. *Democratic Reason: Politics, Collective Intelligence, and the Rule of the Many*. Princeton University Press, Princeton, NJ.

MacKenzie, M.K., Warren, M.E., 2012. Two trust-based uses of minipublics in democratic systems. In: Parkinson, J., Mansbridge, J. (Eds.), Deliberative Systems: Deliberative Democracy at the Large Scale. Cambridge University Press, Cambridge, pp. 95–124.

Mansbridge, J., 2006. Conflict and self-interest in deliberation. In: Besson, S., Marti, J.L. (Eds.), Deliberative Democracy and its Discontent. Ashgate, Aldershot, pp. 107–132.

Mansbridge, J., Bohman, J., Chambers, S., Estlund, D., Follesdal, A., Fung, A., et al., 2010. The place of self-interest and the role of power in deliberative democracy. J. Polit. Philos. 18 (1), 64–100.

Mansbridge, J., Bohman, J., Chambers, S., Christiano, T., Fung, A., Parkinson, J., et al., 2012. A systemic approach to deliberative democracy. In: Parkinson, J., Mansbridge, J. (Eds.), Deliberative Systems: Deliberative Democracy at the Large Scale. Cambridge University Press, Cambridge, pp. 1–26.

Meadowcroft, J., 2011. Sustainable development. In: Bevir, M. (Ed.), The Sage Handbook of Governance. Sage, London, pp. 535–551.

National Research Council, 2008. *Public Participation in Environmental Assessment and Decision Making*. The National Academies Press, Washington, DC.

Öberg, P.O., 2016. Deliberation. In: Ansell, C., Torfing, J. (Eds.), Handbook on Theories of Governance. Edward Elgar, Cheltenham, pp. 179–187.

Organization for Economic Co-operation and Development (OECD), 2012. *OECD Environmental Outlook to 2050: The Consequences of Inaction*. OECD, Paris.

Parkinson, J., 2006. *Deliberating in the Real World: Problems of Legitimacy in Deliberative Democracy*. Oxford University Press, Oxford.

Parkinson, J., 2012. Democratizing deliberative systems. In: Parkinson, J., Mansbridge, J. (Eds.), Deliberative Systems: Deliberative Democracy at the Large Scale. Cambridge University Press, Cambridge, pp. 151–172.

Parkinson, J., Mansbridge, J. (Eds.), 2012. Deliberative Systems: Deliberative Democracy at the Large Scale. Cambridge University Press, Cambridge.

Pogge, T., 2013. Concluding reflections. In: Brock, G. (Ed.), Cosmopolitanism versus Non-Cosmopolitanism. Oxford University Press, Oxford, pp. 294–320.

Quick, K.S., Bryson, J.M., 2016. Public participation. In: Ansell, C., Torfing, J. (Eds.), Handbook on Theories of Governance. Edward Elgar, Cheltenham, pp. 158–169.

Risse, T., 2010. *A Community of Europeans? Transnational Identities and Public Spheres*. Cornell University Press, Ithaca, NY.

Scrase, I., MacKerron, G. (Eds.), 2009. Energy for the Future: A New Agenda. Palgrave Macmillan, Basingstoke.

Steffek, J., 2010. Public accountability and the public sphere of international governance. Ethics Int. Aff. 24 (1), 45–67.

Strunz, S., Gawel, E., Lehmann, P., 2015. Towards a general "Europeanization" of EU member states' energy policies? Econ. Energy Environ. Policy 4 (2), 143–159.

Tews, K., 2015. Europeanization of energy and climate policy: the struggle between competing ideas of coordinating energy transition. J. Environ. Dev. 24 (3), 267–291.

Wissenschaftlicher Beirat der Bundesregierung Globale Umweltveränderungen (WBGU), 2011. *World in Transition: A Social Contract for Sustainability*. WBGU, Berlin.

Zürn, M., 2010. Global governance as multi-level governance. In: Enderlein, H., Wälti, S., Zürn, M. (Eds.), Handbook on Multi-level Governance. Edward Elgar, Cheltenham, pp. 80–102.

PART 2

Case studies

CHAPTER

7

The Kopernikus Project ENavi: linking science, business, and civil society

Piet Sellke[1], Matthias Bergmann[2], Marion Dreyer[1], Oskar Marg[2] and Steffi Ober[3]

[1]Dialogik Non-profit Institute for Communication and cooperation research, Stuttgart, Germany [2]Institute for social-ecological Research, Frankfurt, Germany [3]Zivilgesellschaftliche Plattform Forschungswende, Berlin, Germany

OUTLINE

The Role of Public Participation in Energy Transitions
DOI: https://doi.org/10.1016/B978-0-12-819515-4.00007-6

7.1 Introduction

Germany's Energiewende (energy transition) encompasses not only the phasing out of nuclear energy and coal industry. This transition comprises a deep systemic change toward renewable energy sources as well as toward sustainable consumption patterns in all sectors, whether that is transportation, heat, power grid infrastructures, or consumption patterns. The decision-making process for leaving nuclear and coal energy behind was shaped by complexity due to the ripple effects expected from shutting down two big industry sectors in one of the world's biggest economies. Many of the resulting questions are unsolved as of today. However, the issue is not the restructuring of two industries and their ripple effects alone. To achieve a CO_2 neutral Germany in 2050, fundamental changes in the traffic sector, the building sector, and energy infrastructures are necessary.

An endeavor such as the energy transition creates a tremendous potential of new innovative business models, technologies, and services. But above all, it creates great uncertainty as to what infrastructures, incentives, and regulations are required for exploiting this potential. Moreover, while the energy transition offers potential for some actors, it can lead to losses for other actors, which means that the acceptance of the energy transition can be low in some parts of the population. The decision to phase out coal industries is a challenge as a whole, however a special kind of challenge for the employees working in the coal industry and for the coalmining areas. Although renewable energies are largely welcomed by the public, the picture changes in many of those communities in which wind parks are built or going to be built concretely. And while the majority supports sustainable transportation, only few are willing to change their behavioral patterns.

How to meet these challenges? The involvement of different stakeholders in the decision-making and implementation process at different levels is often seen as part of the solution. However, such participative processes are not only highly challenging but may turn out even counter-productive, especially when many crucial decisions have already been made in advance, or when there are no flanking debates in politics and society regarding underlying assumptions, strategic considerations, and promising innovations. This is where the research project *Energy Transition Navigation System* (ENavi for short) with its transdisciplinary research approach comes into play. ENavi is funded by the German Federal Ministry of Education and Research (BMBF) for a period of about 3 years (October 2016–December 2019). An inter- and transdisciplinary approach including a close cooperation with practice partners such as civil society organizations or regional experts was part of the funding conditions set out in the BMBF's funding announcement.[1]

In the present chapter, we will outline essential processes through which a transdisciplinary approach has been implemented in the ENavi project and reflect these processes and the conditions and challenges of implementing them against an ideal-typical concept of transdisciplinary research.

The structure of the chapter is as follows: we start with sketching an ideal-typical process of transdisciplinary research by setting out main design principles of such a process. For this, we draw on the work of prominent scholars in the field of transdisciplinary research

[1] https://www.bmbf.de/foerderungen/bekanntmachung-1084.html

(Section 7.2). We then describe the process-related transdisciplinary setup of ENavi (Section 7.3) and discuss the way in which the set-up was implemented in the project with reflections on the limitations and achievements in implementing this approach (Section 7.4). On the basis of this account, we specify some lessons learned with a view to future transdisciplinary energy research. The chapter finishes with a brief summary of the main points made.

7.2 Transdisciplinary research and transdisciplinary dialog

Transdisciplinary research and the challenges of sustainability are inherently intertwined with each other. As indicated above, societal changes such as the energy transition toward sustainable energy consumption are *complex, uncertain and ambiguous* (cf. Sellke and Renn, 2010). *Complex*, as the relationship between two variables are often hard to detect, due to for example intervening variables; *uncertain*, as effects might be stochastic, rare (and thus hard to take into account), or simply unknown; *ambiguous*, because regardless of the scientific expertise provided for a certain question, due to different interests or beliefs results of science as well as the methods and the framing of science can be questioned and actually are a matter of debate. What effects to look at is often a normative decision, not a strictly scientific one.

Two processes can be distinguished as to how these phenomena of complexity, uncertainty, and ambiguity are generally dealt with (Lang et al., 2012): the primacy of science and the primacy of practice. Whereas the primacy of science uses lay-persons input to scientific results, the primacy of practice provides decision support (Lang et al., 2012: 26). The concern of science applying this primacy is the loss of credibility (e.g., reliability, validity, further methodological aspects), whereas the concern of practitioners is the loss of the relevance of results (cf Lang et al., 2012). Staying in one of these two spheres does not lead to greater integration to work on complexity, uncertainty, and/or ambiguity. It rather leads to an emphasis on *one* of these trains of thought and hence to shortcomings in either credibility or relevance of results.

Transdisciplinary research (and dialog) widens the lens and integrates the two positions outlined before. Lang et al. (2012: 26) define transdisciplinarity as "[...] a reflexive, integrative method-driven scientific principle aiming at the solution or transition of societal problems and by differentiating and integrating knowledge from various scientific and societal bodies of knowledge." This definition integrates different requirements, as Lang et al. (2012) point out. Transdisciplinary research is supposed to deal with societally relevant problems, it enables mutual learning processes between researchers of different fields *as well as* actors from outside the research field, often referred to as experts of practice or practitioners,[2] and

[2] We use the term "practitioners" in this chapter. Practitioners have expertise as regards the opportunities, requirements, challenges, and implementation conditions of the energy transition on the basis of their own field of activity. This expertise complements the research expertise of academic researchers in transdisciplinary research on the energy transition (see Defila and Di Giulio 2016: 17). Practitioners have their own interests, preferences, and perspectives, which are influenced by their particular expertise. They are organized actors and differ in this regard from the individual citizen. Examples of practitioners as regards the energy transition are business associations, digital start-ups, trade unions, environmental organizations, consumer organizations, and public utilities.

it is aiming at creating socially robust knowledge that is solution oriented. This definition and these requirements require a dialog agenda that goes far beyond common participatory exercises. As pointed out above, whereas participation is at the most the idea of shared decision-making, transdisciplinary research (and dialog) points out the *mutual* learning experience between academia and practitioners.

The actual integrative work is the core and main challenge of the transdisciplinary research process and starts from the outset. That is, a shared problem framing at the very beginning of any transdisciplinary process is the first crucial task to be jointly performed by researchers and practitioners. Establishing a shared understanding of the problem to be dealt with in the project is more than a simple challenge: different normative assumptions of what is valued, different schools of thought and different agendas regarding the matter in question are coming into play, and trust as well as (process and content) transparency are two important variables to enable participants to take part in this stage of shared research, as all actors have to leave their well-known comfort zone.

A second step following problem framing is the definition of a shared research agenda. The societal problems, giving voice to by practitioners, and the scientific problems are supposed to become one shared research agenda, ideally represented through a diverse team including researchers and practitioners, acting as coowners of this agenda. In its actual research phase, transdisciplinarity means the cocreation of solution-oriented transferable knowledge (Lang et al., 2012), situated in both the societal dialog and the scientific dialog. The last phase of transdisciplinary research is supposed to produce *useful* results for society and *relevant* results for science.

Lang et al. (2012) differentiate several design principles for an ideal-typical transdisciplinary research process. The following table gives an overview of these design principles in each phase of the research process.

The following section provides an overview of the main transdisciplinary process elements employed within ENavi including process adjustments found necessary in the course of the project. In Section 7.4, we will come back to the design principles of an ideal-typical transdisciplinary process (see Table 7.1) and contrast these principles with the project's experiences.

7.3 The transdisciplinary setup within the ENavi project

The main objective of the ENavi project is to develop policy interventions for transforming Germany's existing energy system into one based on renewable energy sources and approaching carbon neutrality. This future system shall offer an energy supply which, in accordance with the "Energy concept 2050" of the German federal government, is reliable, economically viable, environmentally friendly, and socially acceptable. In ENavi, science, business, and civil society organizations are working together and exchanging with political actors for developing options for how this can be achieved. The main outcome that ENavi seeks to produce is a navigation tool that researchers jointly with practitioners can use to gage the effects and side effects of various measures

TABLE 7.1 Design principles of transdisciplinary research.

Phase	Design principle
A: Collaborative problem framing and building a collaborative research team	Building of a collaborative research team
	Create joint understanding and definition of the problem to be addressed
	Definition of research object, research questions, and success criteria
	Design of a methodological framework for collaborative knowledge production/integration
B: Cocreation of solution-oriented and transferable knowledge through collaborative research	Assign appropriate roles for practitioners and researchers
	Apply and adjust integrative research methods for knowledge generation and integration
C: (Re-)integrating and applying the cocreated knowledge	Two-dimensional integration
	Generate targeted products for both parties
	Evaluate scientific and societal impact
Cross-cutting principles	Facilitate continuous formative evaluation
	Mitigate conflict constellations
	Enhance capabilities and interest in participation

Own illustration, content adapted from Lang, D., Wiek, A., Bergmann, M., Stauffacher, M., Martens, P., Moll, P., et al., 2012. Transdisciplinary research in sustainability science: practice, principles, and challenges. Sustain. Sci. 7 (Suppl. 1), 30.

in advance.[3] The project understands the energy transition as a process of broad societal change and links scientific analyses to political and social requirements by taking an inter- and transdisciplinary approach. ENavi is a large collaborative project bringing together 60 associated partners and 20 practitioners.[4] Its interdisciplinary approach is represented by 195 research projects developed in 14 work packages (subprojects). Each work package focuses on a specific angle to contribute to reaching the abovementioned objective of ENavi, that is, development of promising policy interventions for the German energy transition. These angles include the perspective of law, digitalization, systems modeling, social and individual behaviors, and furthermore. Whereas the whole project is supposed to work according to the principles of inter- and transdisciplinarity, one specific working package is in charge of providing processes, structures and support for implementing the transdisciplinary approach (the "team transdisciplinarity" for short).

[3] See the ENavi flyer at https://www.kopernikus-projekte.de/lw_resource/datapool/systemfiles/elements/files/8FAD8F72F8457697E0539A695E86621A/live/document/190808_ENavi_Flyer_2019_ENG_final.pdf.

[4] The associated partners include 24 research institutes, 18 university institutes, 3 nongovernmental organizations, 9 companies, 4 local authorities, and 2 regional authorities. The competence partners are different types of practitioners as regard the energy transition (see the ENavi flyer and footnote 2).

ENavi has established several dialog settings to accomplish the transdisciplinary tasks. The three main dialog settings shall be singled out here. First, practitioners involved with relevant questions and problems in the context of the energy transition are grouped and included as "teams of competence."[5] The expertise these individuals bring in is relevant to societal questions regarding the energy transition, mostly from a company or economy-oriented perspective, however also from a perspective of civil society organizations. A second dialog setting concerns the perspectives and concerns of employees. The group here is composed of human resources managers and members of works councils (works councils group for short). A third dialog setting regards the political feasibility of policy options for the energy transition. The respective group consists of Members of Parliament of the German Bundestag as well as the European Parliament. Besides these permanent groups, which meet (separately or combined) with ENavi researchers regularly throughout the project in *practice–science–dialogs*, the process design encompassed semistructured interviews with actors from organized civil society associations as well as political actors from a local level, which were carried out in an early phase of the project.

7.4 Transdisciplinarity in practice: implementation in ENavi

Corresponding to the design principles laid out in Section 7.2, we begin discussing the processes of transdisciplinary research within ENavi with the first phase, the collaborative problem framing, and building a collaborative research team.

7.4.1 Collaborative problem framing and team building

Soon after ENavi's launch, the project consortium agreed to focus the research work on three focal topics to pursue the project's goals effectively: transformation of the electricity system; transformation of the heating sector; and transformation of the transport sector. The design of the transdisciplinary process had to be adjusted in response to this decision: this was done by establishing three "teams of competence" (one for each focal topic, see footnote 3) and, at a later point in the project, subgroups of the works councils group (again one for each focal topic), which met periodically to deal with the respective research issues.

The first task to which these teams and subgroups contributed was agenda-setting. "Contribution" means that the teams and groups did not codevelop the research agenda. The agenda had been already largely defined by the project partners (mostly research institutions) in the phase of proposal writing. The later members of the competence teams and works councils group were not included in this preproject phase, mainly due to the fact that the funding scheme did not include a financed codesign phase. Instead of an ideal-typical codesign of the research agenda, a "catch-up" codesign was carried out in ENavi: the existing research agenda and the main research questions were presented to

[5] There are three so-called *teams of competence*: one is concerned with the transformation of the electricity sector, one is concerned with the transformation of the heating sector, and one is concerned with the transformation of the transport sector (see Section 4.1).

the teams of competence and works councils group, which gave feedback by specifying and further developing these questions and identifying research gaps. This feedback was passed on to the respective research partners. However, it was difficult for them to fully adjust their research programs in response to this feedback, so only peripheral adaptations were made; where the practitioners' concerns and research interests matched the ENavi research agenda, the researchers certainly felt confirmed in their agenda-setting.

According to the design criteria, the team building as a collaborative team was largely a success. The practitioners and scientists alike were eager to discuss different approaches and to learn from different perspectives. Although not all research questions developed by practitioners could be integrated, and although not all research questions within ENavi found the support of the practitioners in terms of relevance, overall the development of a research agenda meeting scientists' and practitioners' concerns was satisfying, as participants stated in event evaluations.

The team building was fruitful not only in the joint work on the agenda, but even more so in the support function the team transdisciplinarity (including the authors of this chapter) exercised among the project partners. The team transdisciplinarity, however, faced the particular challenge to communicate to the actual research consortium the requirements and benefits of transdisciplinary research in regard to the three focal topics. As the funding procedure of ENavi had asked for a detailed research plan of all involved partners long before the project even had started, it appeared challenging for them to engage in activities, which were not part of this plan, as they were often felt as extracurricular—especially in terms of financial resources. According to funding regulations, it is rather difficult to re-distribute grants in different ways than applied for—however, to some degree, this flexibility is a prerequisite for transdisciplinary research.

The definition of the boundary object, as a further design criteria in the first phase of transdisciplinary research, fell in some ways short as the addressees of the project results and the setup of the results itself were not immanently clear and easy to clarify. ENavi aims at developing options on a road(s)map with sustainable, innovative pathways for implementing the energy transition. The navigation tool, which is ENavi's intended main outcome, is initially intended to evaluate scientific findings on the energy transition on the basis of various criteria before translating these findings into options on this road(s)map.[6] If successful, the road(s)map would be of high scientific and practical relevance. Neither the navigation tool nor the road(s)map turned out to be appropriate boundary objects, however, and it appeared that the sheer number of actors involved in ENavi hampered the development of convincing alternatives and agreement on who the recipient of the research work should actually be.

The collaborative framework for knowledge production and integration was build and mostly successful. The team in charge of implementing the transdisciplinary approach achieved a common understanding and further input by practitioners and scientists alike regarding the further methodological proceedings of the project.

[6] See the ENavi flyer at https://www.kopernikus-projekte.de/lw_resource/datapool/systemfiles/elements/files/8FAD8F72F8457697E0539A695E86621A/live/document/190808_ENavi_Flyer_2019_ENG_final.pdf.

The methods of collaborative problem framing were focused on workshop settings between the teams of competence and the project's scientists. As described above, the shared agenda-setting process led to a high degree of interaction between practitioners and scientists. The objective of mutual learning experiences was pursued through instruments such as world café formats, open spaces as well as fishbowl discussions and other formats. The team transdisciplinarity focused on the highest degree of interaction possible within a specified timeframe.

In respect to the task of collaborative problem, framing and building a collaborative research team the following can be summarized:

- Although the research agenda was largely fixed with the start of the project, there was some degree of flexibility on the scientists' side to integrate feedback from practitioners into their research agenda. If it turned out to be impossible due to factual constraints (e.g., resources), this information was transferred to the practitioners. In contrast to the ideal process of defining a joint research agenda *before* the research actually is kicked-off, this setting had some factual shortcomings which were, however, successfully managed in form a catch-up co-design.
- Team building took place among the transdisciplinary research team and the practitioners as well as between the transdisciplinary research team and the scientists of the project. However, team building between scientists and practitioners seemed to be harder to accomplish, mainly due to resource constraints and different understandings of the process itself.
- The design of a methodological framework for collaboration, knowledge production, and integration proved to be subject to differing interpretations. The transdisciplinary research team had to put effort in repeatedly fostering understanding for this framework.

7.4.2 Cocreation of solution-oriented and transferable knowledge through collaborative research

The team transdisciplinarity formed a strong professional connection to the practitioners, which resulted in moderate-to-high participation rates of the teams of competence and the works councils group in the course of the project.[7] The team transdisciplinarity was thereby focused on providing the practitioners with a clear mandate in all stages of the process, that is, the goals had to be always transparent and meaningful if the teams of competence were to get engaged in meetings.

A clear mandate (and thus clear and appropriate roles) is not always easy to establish. Practitioners will join a research process, such as ENavi, usually with specific interests and objectives. Actors from different areas, such as organized civil society and industry, bring different perspectives together, which often enough follow different logics in their goal attainment. However, the teams of competence and the works councils group were led by the team transdisciplinarity through the process with a clear picture of what could happen at what

[7] As it is typical for longer-term dialog processes, the participation rates decreased in the course of time but remained at a satisfactory level.

stage, thus never over- or underestimating the mandate and the tasks ahead. This transparency and clarity of the process created an atmosphere of fairness and mutual understanding, although the process itself had its shortcomings as described in the preceding section.

For the ENavi scientists, it turned out a significant challenge that it had been missed to foresee a separate budget for transdisciplinary work. For this reason, their cooperation with the practitioners had to be driven above all by individual commitment. Furthermore, transdisciplinary research is rarely practiced in large research projects with as many partners as in ENavi. The institutions taking part in the project *had to* follow their own scientific agenda and logic, as their individual evaluation in terms of fulfillment of grant contracts was dependent on achieving their specific scientific objectives, which were not part of the shared agenda-setting process. Although in all scientific arenas, there is a shared understanding that stakeholder engagement is necessary, there is also a shared expectation that stakeholder engagement does make the research process more complicated and does not automatically foster the scientific quality. However, some scientists started to ask actively for feedback from practitioners to enhance their scientific approach.

The adaption of research methods for knowledge generation and integration, in turn, was only minimally responsive to the needs of the transdisciplinary process itself. Due to the project setup, flexibility to integrate *new* research methods was rather low. However, there were satellite solutions in which a research team and practitioners were collaborating in a new way to solve specific subquestions of the research agenda, which was a very fruitful process.

To sum up:

- Appropriate roles were defined for practitioners, who were always confronted with a clear mandate and high transparency about the flexibility of the process. On the other side, the assignment of appropriate and first of all clear roles on the scientific side was more difficult to accomplish, as the scientific teams had to follow a predefined protocol regarding their funding.
- Integrative research methods were developed through satellite solutions between groups of scientists and practitioners (e.g., in the mobility sector), however not for the project as a whole—which would have been a challenge even in an interdisciplinary setting in regard to the high number of partners.

7.4.3 Reintegration and applying the cocreated knowledge

The reintegration and application of cocreated knowledge were not a core intention of the ENavi transdisciplinary process. Instead, the process of evaluating the scientific results according to principles that are socially relevant—such as legitimacy, legality, economically feasible, ecologically meaningful, and ethically sound—provided the grounds of reintegrating the scientific results into the dialog with practitioners.

The teams of competence, the works councils group, and some experts from civil society organizations worked together with scientists—with the abovementioned limitations—in creating measures to achieve the energy transition in the respective field. These measures were tied together to policy packages, and these packages in turn were evaluated according to the mentioned principles. These evaluations were the subject of reflective dialogs with the teams of competence, the works councils group, and further practitioners.

The results of the scientific evaluations needed to be reshaped to become products suitable for science on the one side and practitioners on the other side. Although the scientific discussions and the practitioners' discussions did not show any differences in their level of expertise, thoroughness, and integration of different knowledge sources, the specific interpretation and conclusions drawn from the results by scientists on the one side and practitioners on the other side differed to some degree and followed the before mentioned aspects of *usefulness* as well as *relevance*.

Further integrative work is still to come, specifically the evaluation of scientific and societal impact. As of today, the evaluation of the *transdisciplinary process* itself will take place in a conference/workshop setting with scientists and stakeholders to reflect on the shared work process (previously there has already been an internal evaluation of the transdisciplinary process in ENavi conducted by the team transdisciplinarity, see the next section).

7.4.4 Cross-cutting principles

The team transdisciplinarity put an emphasis on the cross-cutting principles in different ways. First of all, the team included a research organization, which carried out a formative evaluation of the inter- and transdisciplinary process. Team members of this organization participated at most of the practice–science–dialogs and thus received a very thorough inside view of the core collaborative work between scientists and practitioners. The evaluation was generated based on the findings of structured interviews with representatives of all ENavi work packages, from written reports, from participatory observation in many stakeholder workshops and meetings between scientists and teams of competence as well as from other observations through the whole process. The findings from the evaluation provided a helpful and crucial feedback to the team transdisciplinarity itself and to the project as a whole about the state, progress, and quality of the transdisciplinary process.

Throughout the project, mitigating conflict constellations was mainly focused on scarce resources on the scientists' side (time and money) to participate in the transdisciplinary process. Although scientists were eager to be part of the process, still there were factual constraints to this. Thus the transdisciplinary team tried to mitigate conflicts arising from these constellations.

Furthermore, the transdisciplinary setup of the project as not an ideal transdisciplinary process created some dissonance especially in the beginning of the project. It did result in some confusion that agenda-setting workshops could only set *parts* of an agenda or comment on an existing agenda. These shortcomings, which resulted from the specific funding procedure, could be overcome by the transdisciplinary team and eventually yielded acceptance among all partners.

In a project that runs for 3 years, all involved parties need not only to be interested in the transdisciplinary research but also to really see a surplus in this kind of research. That is, to say, it was an ongoing exercise to point out the value of creating solutions together—instead of "only" informing science about stakeholders' views. The understanding of the opportunities coming with transdisciplinary research fostered the participation rate. Especially the teams of competence, which were invited twice a year at least, showed steady participation rates. This is even more noteworthy as any workshop setting in this context is in a very competitive situation with a multitude of further workshops happening at the same time from different organizations.

The capability to participate, which comprises the specific methods of participating, was in an ongoing review process throughout the project. The transdisciplinary research team created different formats of interaction, whether coming from design thinking to more classical approaches of facilitation to rather new interactive formats. Participants from science and stakeholders felt, so the feedback, integrated in a fair and transparent procedure. These aspects played a crucial role in the formative evaluation of the transdisciplinary process as well.

7.5 Lessons learned

The process of transdisciplinary research within ENavi does not follow the ideal, textbook approach. Some of the factors causing deviations from the ideal-typical approach were due to specific formal project constraints, which more than often are not within the research team's discretion. However, despite these obvious constraints, within ENavi, a successful transdisciplinary process could be established. Following, we will summarize the main lessons learned from the applied procedures.

7.5.1 Selection of practitioners and integration

The selection of practitioners is a crucial and time-consuming process. This is well known among scholars of participation and transdisciplinarity and has been confirmed in ENavi. It is crucial, as—even if there is some flexibility in the later process through different formats—the group of practitioners needs to represent key societal issues. Looking at a complex issue, such as the energy transition, there is literally no type of practitioner *not* affected. Thus to establish a fair and transparent procedure of inclusion is of fundamental importance for the success of the transdisciplinary approach.

Inclusion, however, does not mean that *all* practitioners need to be part of the process, and the one that do not all the time (but continuously). What it does mean though is that all practitioners know why they are included in the process or not.

The integration of practitioners results in a double lesson: First, not all relevant practitioners will take part in the process, and it needs to be clearly communicated and planned how to deal with the lack of relevant practitioners. Second, the continuous integration of practitioners over the project's timespan needs a very close networking process to show the surplus of the transdisciplinary process to practitioners and also scientists. Transdisciplinary research does mean leaving the comfort zone to some degree, and there needs to be a clear incentive to do that.

7.5.2 Transdisciplinarity as the project's DNA

Transdisciplinarity must become part of the research project's DNA. Transdisciplinarity is more than integration of practitioners—and it is often confused with exactly that.

The research plan needs to follow the transdisciplinary approach instead of the individual scientific institutional logic.

To achieve this, it is also but not alone about creating awareness among the involved scientists and practitioners of what is understood by transdisciplinarity. It is also about funding procedures that allow certain settings, enabling scientific institutions to work together in a different logic than usually applicable in most funding schemes.

The lesson learned for the ENavi project thereby is that it is of crucial importance to take into account all the barriers created by certain funding schemes. These barriers, often resulting from formal law, need to be addressed upfront and planned in.

Resulting from the funding scheme and the institutional logic is often an individual set of attitudes within science that does not promote the group effort of transdisciplinarity in the first place. In theory, there are two ways to deal with this issue: either defining individual scientific incentives *within* the transdisciplinary process or making individual scientific incentives *independent* of the transdisciplinary process. Either way it should be avoided that individual incentives are working *against* the transdisciplinary process.

Besides incentives, the authors experienced a steep learning curve in dealing with transdisciplinary settings among all partners (including the authors). A lesson learned here is to do capacity building for transdisciplinarity in a focused manner. Whether that is summer schools for scientists or individual settings, it is of crucial importance that scientists understand the merits of transdisciplinarity.

The same is true for practitioners: in principle as a practitioner one has to choose between cooperation in transdisciplinary research and lobbying on one's own behalf. Transdisciplinarity needs to be the better choice to be chosen. Why? What is the surplus for stakeholders? Better understanding of complexity and ambiguity for relevant societal processes? First hands Insights from excellent research?

To implement transdisciplinarity as the project's DNA means that a crucial lesson learned of ENavi is that being "a little transdisciplinary" does not work. The costs of engaging in continuous dialog formats in that case are higher than the wins.

7.5.3 Coordination and project size

ENavi is outstanding in the implementation of transdisciplinary research with 60 associated partners and 20 competence partners. The high number of partners calls for a very close and stringent coordination, and it is questionable how efficient that can be done.

This is not to say that there are an absolute number of partners one should not exceed if doing transdisciplinary research. However, it is an inevitable effect that a high number of partners call for rather separate silo-work, instead of integrated cross-functional bubbles. The latter is needed for transdisciplinary research, an insight that is known since years. "Projects that first attend solely to discipline-specific sub-tasks, with the intention of initiating an integration process at the end of the project, often find themselves struggling with a problem: the connections needed among the partial results individually produced by the sub-projects are difficult if not impossible to establish" (Bergmann et al., 2012: 116). A problem-oriented (and not a discipline oriented) organization of a transdisciplinary research process and problem-oriented research questions as well are of highest relevance for integrated problem solutions. The question must be asked whether this degree of integrated research can be achieved at all in a network with such a large number and diversity of actors.

Cross-functional bubbles define their achievements in the problem-solving capacity in the scientific and societal sense. Thus instead of engaging scientists and practitioners in several large dialog formats, it might be more successful to build cross-functional bubbles concerned with specific questions, thus creating smaller transdisciplinary teams. These teams can have a different pace and be rather self-organized, besides the fundamental transdisciplinary principles.

7.5.4 Openness in the agenda-setting

The agenda-setting process is a crucial stage within transdisciplinary research. The limitations due to the funding procedure within ENavi were discussed above already. One can assume that many projects have to deal with similar constraints, so it might help to draw some lessons learned for agenda-setting processes under certain constraints.

Ideally, the agenda-setting among all scientific and practice partners takes place before the research process is kicked-off at all. If possible, this situation should always be preferred as it ensures complete openness of the process. One variant of this could be for the funding bodies to grant and finance a preliminary phase to the research team, in which scientific and societal actors first jointly negotiate the research questions and objectives. This means that the sponsors must endure the fact that the project is started with knowledge of the relevant social problem, but that the exact agenda of the research tasks still has to be worked out. This would provide an authentic codesign.

The second best way is to enrich and comment on the predefined agenda established by the scientific teams. We call that "catch-up transdisciplinarity." Lessons the authors learned in this regard are that the direct interaction between practitioners and scientists in the second case is crucial, as often differences in research agendas result from different understandings of methods, definitions, and procedures. Thus although the research team had to propose a definite research agenda to the funding organizations, still some flexibility in adjusting and incorporating practitioners' needs is possible.

7.5.5 Integration of practitioners is not one-size-fits-all

Although being in the same context throughout the project, the authors often experienced that no dialog format is alike. The needs of a specific situation, a specific set of practitioners, a certain point in the process, etc., ask for a tailored design.

The lessons learned in that regard comprise also the fact that tailoring the dialog design to a certain situation cannot adequately be planned for the whole project in detail in advance and once but instead needs to be flexible enough to react to real-time changes instantly. Changes in the design might come from specific topics (e.g., level of complexity), different types of practitioners (e.g., different expectations, different resources, and different experiences with dialog formats), different scientists involved (e.g., different experiences with dialog formats), etc.

Integration of practitioners is no trivial endeavor at all, even less designing the appropriate dialog format. It does need experienced dialog experts for doing this; it is much more than coming together and discuss. In fact, a dialog format gone wrong can spoil the whole process in the project as credibility is lost.

7.5.6 Transdisciplinary research team

Although we argued that transdisciplinarity must be part of the project's DNA to work, it also does need a team dedicated only with implementing transdisciplinary designs. This team, and that certainly is a lesson learned in regard to ENavi, must act as "transdisciplinarity agents." In everything that is planned and done in the project, the transdisciplinarity agents need to take care that the transdisciplinary idea is followed—or at least clearly communicated where it is not.

The transdisciplinarity agents are the junction between practitioners and the scientists. They need to translate information, mediate different expectations, create feedback and learnings for all, push innovative ideas for dialog formats, and feel responsible for the process.

All involved partners need to accept the role of the transdisciplinarity agents. The agents need to be closely coupled with the overall project coordination as the integration of different research results is part of the transdisciplinary idea.

Experience from analyzing and accompanying numerous transdisciplinary research projects also shows, however, that in smaller projects with up to 15 or 20 partners, it is not always advisable to delegate responsibility for transdisciplinary methods and processes to one or more persons specifically being accountable for them. Here, it is often more purposeful if all the partners in the team see themselves as responsible for this and are led by a competent overall leadership.

7.6 Summary

The ENavi project aims at delivering socially feasible, economically sound, ethically just, and environmentally sound solutions for implementing the energy transition in Germany. This chapter provided an overview on the transdisciplinary approach ENavi employed.

After setting out principles of an ideal-typical transdisciplinary process, in which generally speaking transparency and fairness of the process are major variables, we discussed the importance of the mutual learning experience, in which science is supposed to create *relevant* results and society *useful* results. The relationship between these two can be tense at times, and we reflected on our experiences on aligning these two objectives.

A central conclusion is that ENavi did not follow an ideal-typical process of transdisciplinary research, however, still managed the challenge of transdisciplinarity through bringing in flexible approaches without leaving the fundamental principles of transdisciplinary research. Often enough funding schemes and research realities will make the implementation of an ideal process difficult if not impossible, but this fact should not be a reason (or justification) to not applying transdisciplinary research methods. Besides many lessons we draw, one central one is that all involved actors were able to accept a learning curve and be part of the transdisciplinary endeavor.

Nevertheless, aiming at the ideal process already in the application phase of a project is a clear recommendation. To make transdisciplinarity the project's DNA, it must be woven in while applying for the funding scheme. The openness on the funding institution's side for this methodological approach, obviously, is a prerogative.

Acknowledgement

This chapter is an outcome of the Kopernikus-project ENavi (Energy Transition Navigation System) sponsored by the German Federal Ministry of Education and Research (subproject J0, reference: 03SFK4J0)

References

Bergmann, M., Jahn, T., Knobloch, T., Krohn, W., Pohl, C., Schramm, E., 2012. Methods for Transdisciplinary Research. A Primer for Practice. Campus Verlag, Frankfurt/New York.

Defila, R., Di Giulio, A., 2016. Transdisziplinär forschen – zwischen Ideal und gelebter Praxis. Campus.

Lang, D., Wiek, A., Bergmann, M., Stauffacher, M., Martens, P., Moll, P., et al., 2012. Transdisciplinary research in sustainability science: practice, principles, and challenges. Sustain. Sci. 7 (Suppl. 1), 25–43.

Sellke, P., Renn, O., 2010. Risk, society and environmental policy: risk governance in a complex world. In: Gross, M., Heinrichs, H. (Eds.), Environmental Sociology: European Perspectives and Interdisciplinary Challenges. Springer, pp. 295–322.

Websites

<https://www.kopernikus-projekte.de/lw_resource/datapool/systemfiles/elements/files/8FAD8F72F8457697E0539A695E86621A/live/document/190808_ENavi_Flyer_2019_ENG_final.pdf>.

CHAPTER

8

Climate change policies designed by stakeholder and public participation

Jörg-Marco Hilpert[1] and Oliver Scheel[2]

[1]Dialogik Non-profit Institute for Communication and cooperation research, Stuttgart, Germany [2]Center for Interdisciplinary Risk and Innovation Studies (ZIRIUS), Stuttgart, Germany

OUTLINE

The Role of Public Participation in Energy Transitions
DOI: https://doi.org/10.1016/B978-0-12-819515-4.00008-8

8.1 Introduction

The government of the German state of Baden-Württemberg launched 2011 a climate protection act, which was supported by an "Integrated Energy and Climate Protection Concept" (IEKK).[1] In the commissioned report for the preparation of the climate protection act, sectoral reduction targets and measures were proposed in order to achieve the climate protection targets in Baden-Württemberg (reduction of the greenhouse gas emissions by 25% by 2020 and by 90% by 2050; reference year 1990). The conceptual basis for the energy and climate policy in Baden-Württemberg—with additional objectives such as reliable supply, cost security, regional value creation, and civic engagement—was the IEKK. The IEKK proposed strategies based on an energy policy scenario and concrete measures that could be used to reach the IEKK targets.

Furthermore, the Council of Ministers has asked the Ministry of the Environment, Climate Protection and the Energy Sector of Baden-Württemberg to carry out a process of *citizen and public participation* in relation to the development of the IEKK. The nonprofit research corporation DIALOGIK supported the Ministry of the Environment, Climate Protection and the Energy Sector in its intention to realize a comprehensive citizen and public participation in the development of the IEKK for Baden-Württemberg, called: "Citizen and Public Participation for the Integrated Energy and Climate Protection Concept Baden-Württemberg" (BEKO).

Energy and climate policy is a complex and long-term issue, which requires a stable economic and socially accepted framework. Therefore the possibility of participation in the design of the IEKK should be made possible. This corresponded with the claim of the state government to strengthen the involvement and participation of the people in shaping the energy transition and in achieving the climate protection goals. Therefore the IEKK was subjected to an early, comprehensive, and open-ended participation process, the BEKO.

This article presents the background of the BEKO, its participatory elements, and their temporal sequence. This article is completed by the first evaluation results including empiric examples for the special benefits of a participation initiative like the BEKO.

8.2 Background

The state government of Baden-Württemberg has adopted a climate protection law with binding targets for reducing greenhouse gas emissions as the central element for the realignment of the energy and climate policy. This law provides the legal framework for climate protection in Baden-Württemberg. In this law the interests of climate protection in the planning and discretionary decisions of the public sector are embedded, as well as tasks, instruments, and responsibilities for the achievement of objectives within the scope of state law competencies.

Based on the climate protection law and a scenario for different sectors (power supply, private households, industry, trade/commerce/services, traffic, public sector, agriculture, and forestry/land use) so-called sector targets were derived from the country's climate

[1] The climate protection act of Baden-Württemberg came into force on July 31, 2013.

protection targets. On this basis the specialized departments of the responsible departments developed a set of strategies and measures (with the help of external experts) and compiled them in the "IEKK working draft" (175 pages). This working draft contained 110 strategies and suggested measures for the achievement of the energy and climate policy objectives of Baden-Württemberg and formed the substantive basis and the starting point for the BEKO.

The achievement of ambitious climate protection goals and the requisite reorganization of the energy supply means a strenuous effort for all social actors. That is why the state government of Baden-Württemberg decided to put the proposals for action from the "IEKK draft work" for public discussion and asked all citizens and stakeholders of Baden-Württemberg for support and participation.

The task of realigning the energy and climate policy of Baden-Württemberg has far-reaching consequences for many areas of life. According to the state government, expertise, experience, and interests of the citizens can make as effective contributions to their design as the organized public. Especially the future-oriented topic of an integrated energy and climate policy is suitable for the participation of all stakeholders, since far-reaching decisions have to be made in order to achieve the climate and energy policy goals. With the intention to examine the IEKK working draft and to provide recommendations for its further development, citizens and organized associations/stakeholders were invited to contribute their ideas and wishes, in order to give the energy and climate policy in the country an equally socially supported conceptual orientation.

In February 2012, the state government of Baden-Württemberg adopted the key elements of a climate protection act that is supported by an energy and climate protection concept (IEKK). The report that was commissioned for preparation of the act proposes sector-related reduction targets and measures to achieve the climate targets in Baden-Württemberg.

8.3 Procedure of the BEKO

Fig. 8.1 shows the "new way" of the BEKO, which was embedded in the traditional concept development process: After the preparation of the IEKK working draft with its strategies and measures to achieve the energy and climate policy objectives (1), the BEKO was carried out as an early and open-ended participation of citizens as well as organized stakeholder groups (2). This additional informal face-to-face participation for organized stakeholder groups took place well ahead of the "traditional" stakeholder hearings. In addition to participation in separate face-to-face events the citizens of Baden-Württemberg had the opportunity to participate online (see also Fig. 8.4). In the next step the recommendations and comments developed within the framework of the BEKO were handed over to the state government, examined by the responsible departments and taken into account in the further development of the IEKK draft. Following the BEKO the formal association hearing of the organized stakeholder groups on the advanced IEKK draft took place (3). After the adoption of the IEKK by the state government (4), the strategies and measures were implemented (5). The implementation is reviewed and measured on an ongoing basis (6) in order to make any necessary adjustments and follow-up (7).

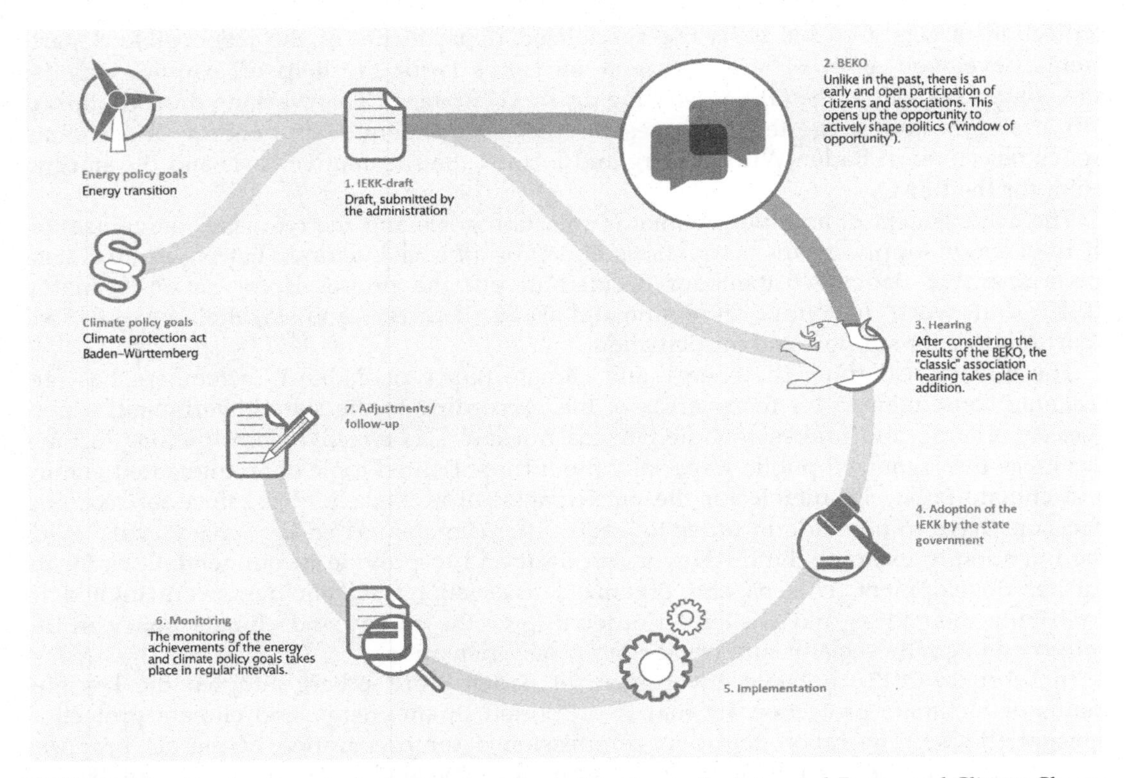

FIGURE 8.1 The "new way" of citizen and public participation in the Integrated Energy and Climate Change Concept (BEKO). Source: *Adapted from Ministry of the Environment, Climate Protection and the Energy Sector Baden-Württemberg, 2012. BEKO Information Brochure "We are breaking new ground! Join us!" <https://um.baden-wuerttemberg.de/fileadmin/redaktion/dateien/Altdaten/202/Anlage_IEKK_und_BEKO.pdf> (accessed 14.07.19.): 6f.*

8.4 Structure and procedural organization of the BEKO

The BEKO had to fulfill a variety of requirements and requests, for example:

1. A broad and inclusive involvement of the public.
2. Enable the development of new ideas.
3. Commenting on the strategies and measures proposed by the state government.
4. The highest possible equal opportunities for participation in the procedure.
5. The activation of knowledge from citizenship and stakeholders.
6. The verification of willingness to approve proposed strategies and measures.
7. With regard to a further development of the IEKK, the recommendations and comments should be verifiable (result orientation).

In order to meet these requirements and requests, different methods were used. The following section deals with the elements within the BEKO in more detail.

8.4.1 Online survey

All citizens in Baden-Württemberg should in principle have the opportunity to participate in the BEKO. On the website http://www.beko.baden-wuerttemberg.de/, they were invited from December 2012 to February 2013 to rate and comment the suggested measures of the IEKK and to make their own suggestions.

Because all visitors had potentially access to the online participation, they were first asked to select the state, respectively the Regierungsbezirk[2] in Baden-Württemberg where they live (Baden-Württemberg is divided into four administrative districts) (see Fig. 8.2).

Thus the probability could be increased that only votes from Baden-Württemberg went into the evaluation. The assessments and comments were then summarized and introduced in the sessions of the roundtables. In addition, the results were made available on the website.

After choosing the state, respectively the administrative district of Baden-Württemberg, the visitors were informed about the evaluation procedure, for example, that the proposed

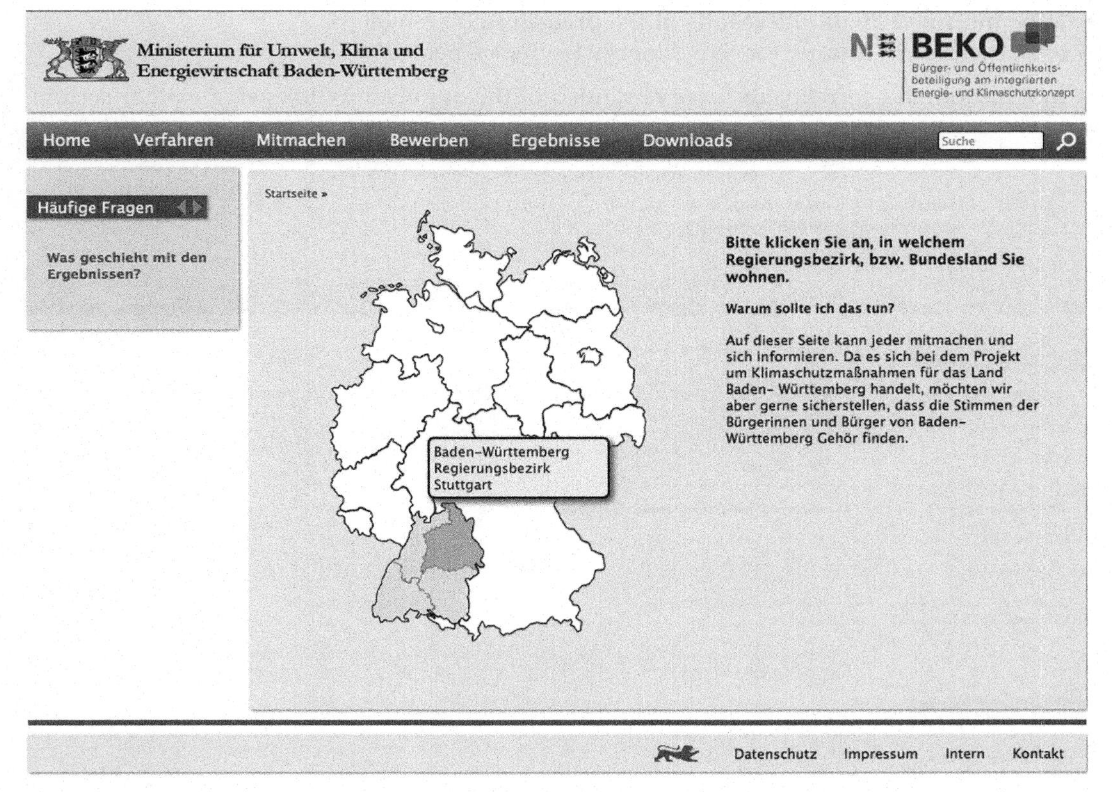

FIGURE 8.2 Screenshot of the start page into the BEKO online participation with selection of the federal state, respectively the administrative district.

[2] Regierungsbezirk is a German administration district below the state level, almost like regional board in the United States or the regional council in Great Britain. Only West-German states with high population density use this administrative unit.

measures for each sector are displayed by random numbers and therefore do not necessarily indicate consecutive numbers. Thus every proposed measure was given a probability of being included in the sample. Visitors were able to choose for each proposed measure on a scale from −4 to +4 whether they would "not recommend at all" (−4) or "highly recommend" (+4) the implementation of this measure to the state government (see Fig. 8.3). Additionally, they could choose "I do not understand the measure" or they could not give no response. There was also the possibility to comment every proposed measure.

After saving a rating, the bottom and the left progress bar changed. The bottom bar indicated what percentage of the proposed measures within a sector have already been processed. Since some of the proposed measures had also been implemented in other sectors, it was possible that the left progress bar also changed in another sector. This visualization should motivate the visitors to evaluate proposals from other sectors as well.

After finishing a sector, visitors had several options: They were able

- to apply for a so-called 5th citizen table,
- to be informed about the results of the procedure by e-mail, or
- to participate in a raffle for five "Energy Savings Checks."

In addition, the participants were reminded of the opportunity to evaluate other sectors.

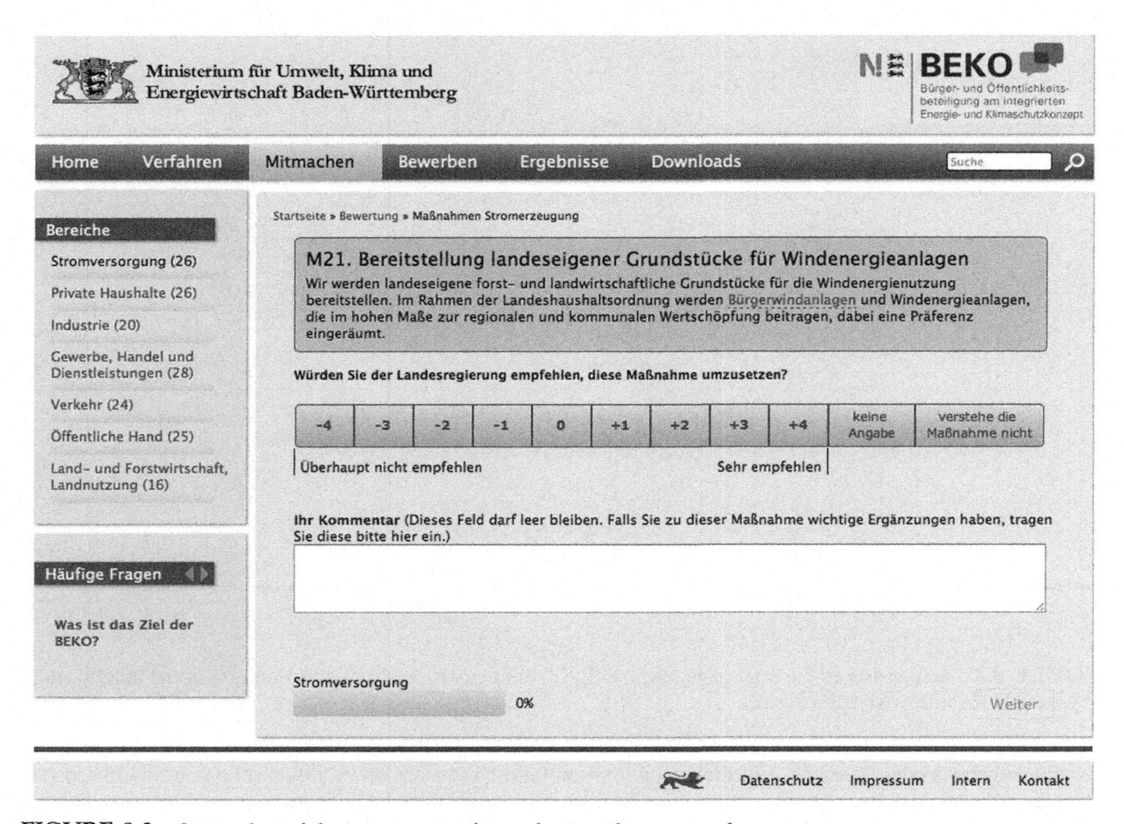

FIGURE 8.3 Screenshot of the input screen for evaluating the proposed measures.

8.4.2 Citizens and stakeholder/association tables

A special feature of the process of the BEKO was that both citizens and representatives of organized associations and interest groups were involved in the process (see Fig. 8.4). The respective "citizen tables" and "stakeholder/association tables" met separately and passed "own" recommendations.

At the end of each meeting, representatives from both tables (e.g., the citizens table "power supply" and the association table "power supply") were brought together in so-called reflection session 1 to discuss the possibility of joint recommendations (on the power supply sector). In a final round ("reflection session 2") representatives from all tables were invited to formulate joint, cross-sectoral recommendations.

In order to facilitate a statewide participation, a citizen table with randomly selected citizens took place in each of the four administrative districts of Baden-Württemberg. Since most of the associations and interest groups are based in Stuttgart, all seven stakeholder/association tables were held in the state capital Stuttgart.

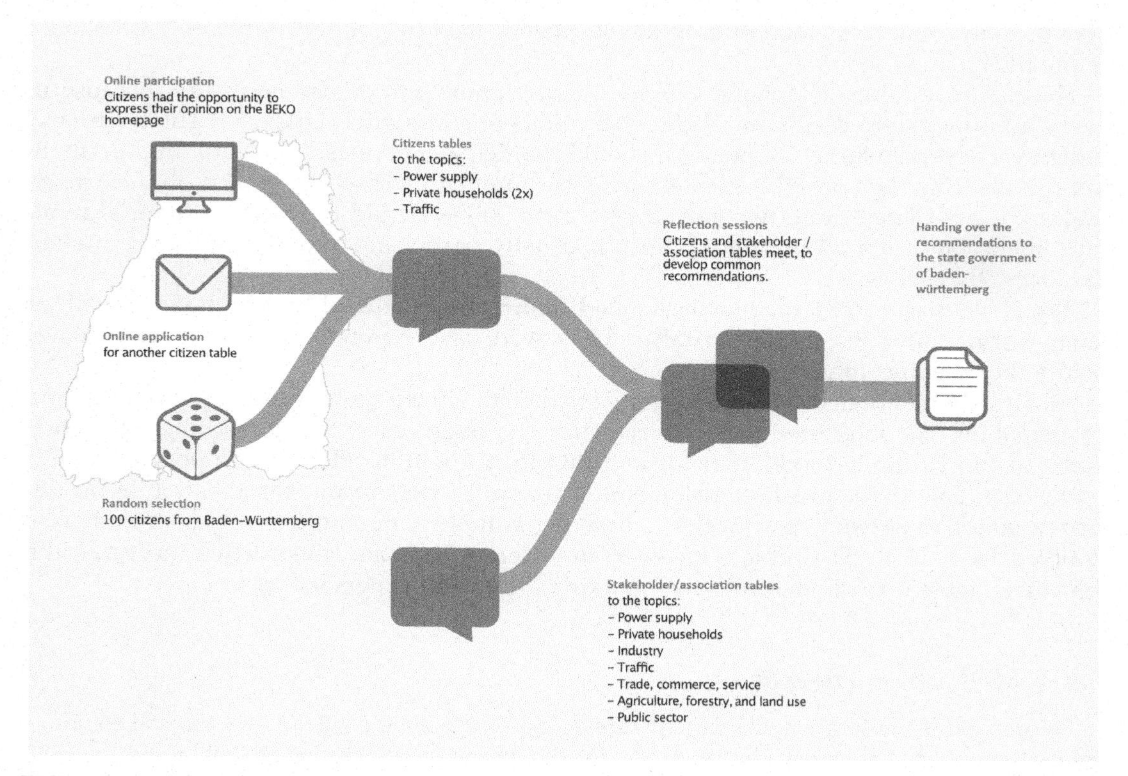

FIGURE 8.4 Process of the civic and public participation for the Integrated Energy And Climate Protection Concept Baden-Württemberg (BEKO). Source: *Adapted from Ministry of the Environment, Climate Protection and the Energy Sector Baden-Württemberg, 2012. BEKO Information Brochure "We are breaking new ground! Join us!" <https://um.baden-wuerttemberg.de/fileadmin/redaktion/dateien/Altdaten/202/Anlage_IEKK_und_BEKO.pdf> (accessed 14.07.19.): 10f.*

Each of the sessions lasted about 3.5 hours and was led by trained moderators. In addition, representatives from the responsible departments were present to provide explanatory background knowledge, if necessary, and to answer open questions directly on-site.

The aim of the sessions was to comment the proposed measures, respectively to formulate recommendations for their completion or further development. The participants could, for example, include the following aspects in their deliberations:

- Are the measures useful or are there others that are more suitable?
- Does a practical implementation of the measure encounter unacceptable hardship?
- Should the proposed measures remain that way or should they be reformulated?

8.4.3 Selection of citizens for the tables

Similar to a representative survey, the participants of the citizen tables were identified by a random telephone sample. By doing so all residents of Baden-Württemberg basically had a comparable chance to participate in a citizen table. Every person called was free to take this opportunity. A total of around 180 citizens took this opportunity and participated in the BEKO.

For such a random telephone sample a considerable effort was necessary, because in every administrative district of Baden-Württemberg (Stuttgart, Karlsruhe, Tübingen, and Freiburg) approximately 25 persons should participate. It was also very important to inform the citizens with the greatest possible attention about the topics of the citizen tables, the associated time (two sessions for every subject), and the offered financial reimbursement. Since not all registered persons usually participate, more people were invited than actually necessary.

The citizens were provided with detailed information material by telephone, e-mail, or letter during/after the invitation talks. They were also reminded of the meeting dates before the meetings took place.

As already mentioned, the participants for the 5th citizen table could apply online for a participation. The interested people were informed in advance that the number of participants would be limited to 25 persons and that the lot will decide, if more people want to participate. Due to the great interest (about 400 applications for the participation in the 5th citizen table), in agreement with the contracting authority, the number of participants was doubled from 25 to 50 people. However, in order to facilitate constructive meetings, the 5th citizen table was divided into two working groups of 25 persons each.

8.4.4 Moderation concept

Beforehand a moderation concept was developed for both the citizen and the stakeholder/association tables. The concept of the tables for the first session included the following:

- Welcoming,
- Overview of the principles of the procedure,
- Brief introductory round of the participants,

- General information on the measures,
- Evaluation on the measures from online participation.

Subsequently, an atmospheric picture of the proposed measures with special need for discussion was developed. For this purpose the participants received blue adhesive dots (corresponding to the number of the proposed measures in their sector). Similar to the online evaluation they were able to rate the suggested measures on a scale of "not recommend at all" (−4) to "highly recommend" (+4). The participants were also able to indicate whether they found the wording of the proposed measures as meaningful enough. In addition, the participants were given the opportunity to emphasize about one third of the proposed measures with red adhesive dots to mark then as "discussion requested."

As next step a discussion of all proposed measures took place, which turned out to be particularly "worthy of discussion" (marked as "discussion requested") (including justifications, explanations, comments, etc.). The moderator designed the discussion between the participants in such a way that each voice was heard, and all participants could express their opinion. In addition, new proposals for measures could be introduced by the participants.

At the end of the first session of meetings, both the participants of the citizen and the stakeholder/association tables were reminded to choose two persons who represent their tables at the reflection session(s). The actual election then took place in the second session of the tables. The second session started with a review of the first session and the evaluations given there. Subsequently the discussion of the proposed measures, which had already been marked in the first session, continued. At the end an outlook was given on the further process of the procedure.

Since the 5th citizen table was not limited to measures for individual sectors, the moderation concept and the procedure differed in the following way: Unlike the other tables, all 110 proposed measures were up for discussion. In order to be able to cope with the given time budget at the beginning of the session an attempt was made to identify those proposed measures on which the further discussion should concentrate: At the beginning the results of the online participation were presented and the moderator proposed to discuss those measures in particular, which were rated rather ambivalently (without clear tendency) in the online participation. The participants also had the opportunity to evaluate the proposed measures in regard to their need for discussion.

8.5 Implementation of the BEKO

8.5.1 Citizen tables

At every of the four citizen tables up to 35 citizens discussed, commented, and supplemented the proposed measures of the IEKK working draft. For participation at the citizen tables no specific prior knowledge was required. Rather it was the concern of the state government of Baden-Württemberg to incorporate the citizens' personal judgment and life experience in the decision on energy and climate policy in Baden-Württemberg.

TABLE 8.1 Citizen tables in the administrative districts of Baden-Württemberg.

Subject	Administrative district
1. Power supply	Karlsruhe
2. Private households	Freiburg
3. Private households	Tübingen
4. Traffic	Stuttgart

TABLE 8.2 Stakeholder/association tables in the districts of Baden-Württemberg.

Subject	Administrative district
1. Power supply	Stuttgart
2. Private households	
3. Industry	
4. Traffic	
5. Trade, commerce, service	
6. Agriculture, forestry, and land use	
7. Public sector	

For reasons of efficiency, in each administrative district of Baden-Württemberg a citizen table on a specific subject area of particular relevance for the citizens was organized (see Table 8.1).

As already mentioned, in addition to the four citizen tables mentioned above, another so-called open topic "5th citizen table" was conducted.

8.5.2 Stakeholder/association tables

In addition to the citizen tables, representatives of organized associations and interest groups from Baden-Württemberg were invited by the Minister of the Environment to so-called stakeholder/association tables. The focus was on organizations from industry and society, for whom it was important to shape the climate and energy policy. This included, for example, associations and interest groups from the fields of industry and trade union, handcraft, trade, transport, agriculture, environmental protection, or consumer protection. The participants of the stakeholder/association tables received the same documents and information as the participants of the citizen tables. The participants of the stakeholder/association tables were also able to discuss, evaluate, modify, or supplement the proposed measures.

However, the thematic offer concerned all 110 proposed measures, which is why the following seven stakeholder/association tables were realized (see Table 8.2).

The tables were also moderated externally, with each table focusing on one of the topics listed in Table 8.2. The moderation concept for the stakeholder/association tables was largely congruent with the moderation concept of the citizen tables. Unlike the citizen tables all stakeholder/association tables were held in Stuttgart, because most of the organizations are based in Stuttgart.

In total, around 120 representatives of associations and interest groups participated in the process.

8.5.3 Reflection sessions

Approximately two weeks after the last session of the citizen and stakeholder/association tables, moderated "reflection sessions" were realized. The purpose of the reflection sessions was to examine the results of the citizens and stakeholder/association tables on similarities and, if possible, to formulate joint recommendations to the state government.

At first reflection sessions were held on topics that were addressed at the citizens as well as at the stakeholder/association tables ("power supply," "private households," and "traffic"). The two elected persons of the respective civil and stakeholder/association tables attended these sessions. In advance, the elected persons were given a comparison of the recommendations of both tables, which were formulated on the same topic. The representatives then discussed these recommendations and tried to formulate shared recommendations.

At the second reflection session the representatives of all citizens and stakeholder/association tables took part (altogether 24 persons). The aim of this final sessions was to develop joint strategy recommendations. Therefore the discussion did not focus on (individual) proposed measures or recommendations as in previous meetings, but the participants worked in the following categories:

1. *Information, education, integration*: Recommendations that are useful for informing the public (e.g., brochures, advisory services, information events, and materials for schools and training).
2. *Coordination, roundtables, joint certification, cooperation between private sponsorship and the public sector*: Recommendations that include proposals for processes, procedures, or cooperation offered between the involved actors (including certification process).
3. *Own measures of the public sector, planning, research funding, and self-commitment*: All recommendations made or at least initiated by the public sector had been summarized in this category. They do not require any special services from third parties, nor do they commit them to doing anything, but bind the public sector to measures that they must take responsibility for and finance them by themselves (e.g., energy-saving measures in public buildings, reform of the company car regulation, changes in public procurement or special research funding).
4. *Infrastructure measures*: This category includes all recommendations that serve to build or change infrastructural services.
5. *Financial incentives or burdens*: All cash benefits of the public sector for projects that are carried out by private bodies (industry, associations, non-governmental organizations,

consumers) were included in this category. This category also included all burdens, such as additional duties or taxes and special financial support programs for consumers.

6. *Regulatory measures*: This category includes all recommendations that limit or prescribe specific behaviors, as well as changes in building and planning regulations, which set new priorities for environmental planning (such as climate protection against nature conservation). In addition, changes in traffic law (e.g., speed limits) had been classified here.
7. *Climate and energy policy in general*: This category included all recommendations on "Climate and energy policy in general."
8. *Procedure of the BEKO*: All recommendations regarding the "BEKO procedure" were assigned to this category.
9. *Reformulations, concretizations, links*: All recommendations to "reformulations, concretizations, links" were added here.
10. *Other recommendations*: This category included all recommendations that could not be classified in any of the other categories.

8.6 Recommendations and comments

In total, 272 recommendations from the citizen tables, 334 recommendations from the stakeholder/association tables, and 145 recommendations from the reflection sessions were formulated within the framework of the BEKO. These recommendations were handed over to the state government of Baden-Württemberg on May 2, 2013, as part of a state press conference. The recommendations had been reviewed by the respective departments regarding their feasibility, effectiveness, and affordability by the public budget and their cost burden.

In addition, further 331 recommendations were derived from the 6.742 comments from the online participation. A list of all 6.742 comments has been published at www.beko.baden-wuerttemberg.de/results.

The state government of Baden-Württemberg has fulfilled its commitment to thoroughly review all recommendations and to incorporate them into an updated IEKK draft, if the result of the examination is positive. If a recommendation could not be followed, for example, because the respective recommendation was not financeable or legally unenforceable, an attempt was made to identify the intention behind the recommendation and to take it into account as far as possible.

Following the review of the BEKO recommendations by the state government an extensive document (more than 400 pages) has been created and published, which contained detailed results of each of the 1.082 recommendations. At the same time the IEKK draft was updated on the basis of the results and other developments that had occurred in the meantime. With this further developed version of the IEKK draft the formal hearing of federations and associations was initiated.

After taking into account the results of this hearing as well as a response from the state parliament the state government has approved the IEKK. After the approval of the IEKK the strategies and measures were started to be implemented. A monitoring ensured that, in case of deviations from the achievement of goals, the strategies and measures could be adapted, and adjustments could be made.

8.7 Evaluation

The BEKO was accompanied by an external evaluation which demonstrated high satisfaction with the process by all participants and a particular appreciation of the inclusion of all major stakeholders.

8.7.1 Challenges and specialties of the results of citizen and public participation

The whole participation process of the BEKO addressed the citizens of Baden-Württemberg using different media. Thus the evaluation, in order to enable a holistic analysis of the participation process, had to include all possible participation formats available to the citizens. In addition to that the evaluation tried to include not only citizen as recipients of the participation but also its initiators, the public administration—namely representatives of the political parties of the parliament of Baden-Württemberg, alongside officials from five departments of the state, that were included into the formulation of possible measures with respect to the climate change that were to be discussed within the BEKO.

Fig. 8.5 gives a schematic overview about the evaluation strategy of the BEKO. As shown the BEKO evaluation included multiple methods from qualitative and quantitative social science to gain a thorough and holistic picture of the ongoing participation process, its output, and its outcomes from inside and outside perspectives.

The mixture of qualitative questionnaires and quantitative interviews exceeds that of other evaluations with respect to the inclusion of further, indirect measures and observations: The online websites of the BEKO not only protocolled the recommendations to every single proposed measure plus the evaluation (on a scale between +4 and −4) given by the participants. Furthermore the usage of the whole website was protocolled and thus

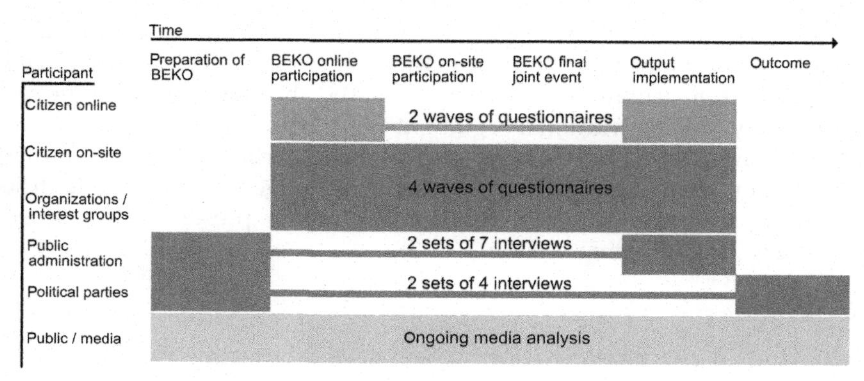

FIGURE 8.5 Evaluation of the BEKO participation (own representation).

information about whether visitors returned to the website, the time spent during a session, and exactly what pages were viewed at all could be used for the evaluation.

Compared to an average participation process, which includes the measure of perception, cognition, and satisfaction of its participants in two points in time (usually pre- and post-participation) significantly more data was generated during this evaluation.

The early inclusion of interviews among officials of the five governmental departments of Baden-Württemberg that created the proposed measures, in particular the IEKK draft, was, for example, such an important additional data source: It opens the possibility to check whether the perception of the participations' purpose was congruent (1) among different groups of the imitators of the participation and (2) between participants and initiators. Such a mutual understanding is the foundation for any successful participation process. This point seems to be obvious, but often enough evaluation programs find that even a miscommunicated (or misunderstood) degree of freedom in the process of generating joint outputs of participation processes leads to frustrated citizens or disillusioned administration officials. Section 8.8.2 will show that the BEKO measured, that a clear communication strategy enables even the ability to foster mutual process learning between stakeholder organizations, citizens, and public administration—a finding of great importance, in times of diminishing satisfaction with democratic institutions.

Other unconventional data gathered in the BEKO evaluation were the "observation" data of participants' behavior on the online webpages during the public phase. These data equal the behavior observation which usually is part of any on-site participation event and a vital part of any evaluation, as it is the best measure on the dynamics of social interaction during the participation process. Participation processes that are conducted online have usually far less social interaction, thus an observation of the single participant and its usage of given information is even more important. Section 8.7.2 will present some explanatory results that show the importance of such additional data sources for an evaluation.

Another fact about public participation is that participation not only fails if imitators and participants misunderstand the goals of a participation process but that several different, willingly participating, societal groups might not trust each other and thus are an obstacle to any communication and negotiation which render a participation doomed from the beginning. In the BEKO two such groups met at the final "reflection session" and the previous surveys showed that there were severe reservations on both sides toward each other: citizens feared to be pushed aside by professional lobbyists, sent in by the co-invited stakeholder organizations or marginalized by being labeled unprofessional. The invited stakeholder representatives on their side feared that the citizen might block any of their perspectives by voicing irrational arguments and fears.

As the evaluation included not only surveys on both participant groups but also before and during the joint discussion phase, these critical dangers could be identified by the BEKO evaluation. How this potential conflict resolved is presented in the explanatory results presented in the following section.

8.7.2 Evaluation of the participation process—explanatory results

This section presents selected results from BEKO evaluation as a full report would need a book of its own. We selected the presented results to underline our statement that the unusual and in-depth data gathering conducted during BEKO participation lead insights into the participation process and these could be repeated in other processes. As not every participation could be fitted to the challenging features of BEKO—the five governmental departments that the 1.082 recommendations were addressed at invested five months of time to generate an updated draft of the IEKK—but at least other participation projects could use the given experimental-gained knowledge to ensure better participation results.

A clear communication of intension and goals of any given participatory process between all stakeholder groups, public, and initiators was a goal of the BEKO from the beginning. To ensure and control its implementation and success all participants received prior to the second respective fourth questionnaire (see red and green parts of evaluation in Fig. 8.5) access to three versions of reports about the output of the BEKO participation:

1. A ten-page listing of all changes in the wording of the measures within the updated IEKK draft for readers interested to check whether their special point of interest was included in the improved IEKK.
2. A 180-page version of the updated IEKK draft that highlighted every change made that recurred to the written recommendations and the spoken ones during the BEKO online phase and any of the on-site events.
3. The complete report, 410 pages in length, that listed all given recommendations together with the complete written reaction of the five governmental departments to each of them. This could be used by the most avid participants to understand the reasoning of public administration.

After offering such custom-tailored information opportunities we included in the post-survey the items presented in Table 8.3 which were replied greatly in favor (all means are on the positive side of the scale from -4 to $+4$) for the BEKO evaluation, warranting the satisfaction of the participants. Even more the participation made it possible to understand the boundaries and limitations when negotiating compromises of the other groups encountered during the events.

As the survey includes also open questions, the evaluation can use direct citations given by the participants to underline this: "*the consideration* [of peoples' comments] *in the IEKK is very promising. I hope the implementation will be successful.*"[3] Such a second look at quantitative data by using qualitative data is one of the special traits of the BEKO evaluation to underline that the communication between initiators and participants, about what the BEKO intended, worked.

A second example from the evaluation analysis is the exact observation of user behavior on the website: On the BEKO website the tracking of time spent on different subpages can be used to check whether the user actually read or worked with the given information or whether they just clicked out of curiosity. Table 8.4 shows the most viewed pages of the BEKO online participation alongside with the time spent on the different pages. In red highlighted are the pages where the (in seven segments sorted analogue to Table 8.2, see Section 8.6.2) BEKO measures were presented and the users could rate and comment each of them.

[3] Participant 78, post-survey, F7; the answer has been translated from the German original by the authors but not altered in any way.

TABLE 8.3 Post participation satisfaction, scale: affirmation of item between +4 and −4 (own representation).

Item (translated into English)	Mean affirmation	*N*
"the participants perspective and recommendations are visible in the updated IEKK draft"	+1.70	105
"I learnt about other participants' perspectives during the discussions in the BEKO events"	+2.92	50[a]
"I gained trust in participation prepossesses in general after being part of the BEKO"	+1.42	105
"I understand that not all recommendations given during the BEKO could be part of the updated IEKK draft"	+2.38	105

[a]*This question was only given to the participants of the on-site events, not the online participants.*

TABLE 8.4 Fourteen most visited pages of the BEKO online participation (own representation).

Rank	Page name[a]	Number of page views	Number of users[b]	Time spent (min:s)
1	/entrance	5911	3759	1:05
2	/join in	3340	2493	0:40
3	/state	2769	2562	0.18
4	/survey/1	2334	1675	4:54
5	/procedure	1733	1346	1:15
6	/apply	1683	1203	1:02
7	/downloads	1588	1095	2:33
8	/survey/2	1556	1223	5.06
9	/survey/5	1012	797	5:55
10	/results	923	708	0:17
11	/survey/7	785	606	3:34
12	/survey/6	677	607	3:16
13	/survey/3	669	590	3:16
14	/procedure/citizen tables	632	476	1:29

[a]*Page names are translated into English to enable reader to understand information presented on the pages.*
[b]*Based on nonrecurring cookies.*

The analysis shows that most users of the website used the majority of their time spent on the website to read the proposed measures against climate change and to rate and comment on them. Compared to all the other information presented, these pages are visited 3–10 times longer—even compared to pages that also demanded input from the users to proceed, like

- /apply, where online participants could apply to be invited to an on-site event, the "5th citizen table" to discuss the measures together with other citizens in person.

TABLE 8.5 Users of the BEKO website separated after residence in German states—self-proclaimed (own representation).

German state	Number of users	Number of measure ratings	Administrative districts of Baden-Württemberg	Number of users	Number of measure ratings
Berlin	96	121	Freiburg	558	14.152
Brandenburg	87	110	Karlsruhe	627	16.819
Bavaria	124	657	Stuttgart	1.159	32.639
Hessia	105	541	Tübingen	574	15.545
Bremen	84	0			
Hamburg	88	120			
Mecklenburg-Vorp.	86	10			
North Rhine-Westphalia	113	269			
Lower Saxony	102	564			
Rheineland-Pfalz	92	361			
Saxony	88	110			
Saxony-Anhalt	87	8			
Schleswig-Holstein	92	111			
Saarland	86	64			
Thuringia	88	4			
Sum Germany	1.418	3.050	Sum Baden-Württemberg	2.918	79.155

- /state, where users had to announce their home state or administrative district within the state of Baden-Württemberg.

As the website used cookies to recognize the users, the "number of users" column might present a slight overestimation of users (for people using different computers or smartphones to access the page or regularly delete cookies) but gives a greatly enhanced estimate of the true sample size (see Table 8.5).

The page "/state" was of special importance as it helps to estimate the range the BEKO participation reached within Germany, especially within the state of Baden-Württemberg, whose population was the goal of the participation. As self-recruiting online formats lack a true guidance who is included, we decided not to force untargeted users out but to check who we reached by offering a small Germany map to click on a state where the user is living (see Fig. 8.2). This strategy proved to be a success: though the information about what the BEKO participation intends reached even without further

TABLE 8.6 Users of BEKO website compared to population (own representation combined with numbers from the statistic office of Baden-Württemberg, December 31, 2018).

Administrative districts of Baden-Württemberg	BEKO user (%)	True population (%)
Freiburg	19.12	20.4
Karlsruhe	21.49	25.3
Stuttgart	39.72	37.4
Tübingen	19.67	16.7

marketing outside of Baden-Württemberg (1.418 users, see Table 8.5), the clear statement that only citizens of Baden-Württemberg are asked to comment on the measures leads to only 3.7% survey entries to be marked as not from the target population. The vast majority of 79.155 measure ratings were direct transferable to the local administration (see Table 8.5).

Acknowledgment: As the page was only presented to users who wanted to access the survey, not the entrance or information pages, the number of users differs from Table 8.3.

Also the page offered a checkup whether the different regions within the state ware well represented within the results of the participation: As Table 8.6 shows the different users of the BEKO website matched very good with the distribution of residents among the administrative districts of Baden-Württemberg (see Table 8.5).

A final example of how the BEKO participation turned out will be given in the case of the potential conflict between stakeholder organization participants and the citizens included in the participation process: But the BEKO turned, due to professional moderation of the participation process for both sides, into a positive and even informative experience: Table 8.7 shows statistical data from the responses to different items circling around the participation process and its enforced rules by the moderation sorted by the participants' group membership. All the means presented are based on the +4 ("totally agree with item presented") to −4 ("totally disagreement with item") scale that was used in all written surveys of BEKO. Both groups showed high interest in the same, fair discussion rules for all participants, the unbiased moderation could foster a fair and just exchange of arguments for both groups and thus create a positive experience.

This also leads to a high satisfaction that

- mentally connected the BEKO participation to good democratic decision processes (see Item 5) and
- lead to an attitude to be open for further events of the same type (see Item 6) among the most participants.

The lower rate of affirmation among the stakeholder organization representatives could be decoded by open questions which lead to precise explanations like this by a participant:

> The participation is made to include people without previous knowledge (important for discussions with laypeople), the moderation tries to lead to unitary statements (good for discussions with laypeople),

TABLE 8.7 Means of responds to items identical posed to citizens and stakeholder representatives after BEKO events (own representation).

No.	Item (translated by the authors for readers' comprehension)	Mean item affirmation citizen	Mean item affirmation stakeholder
1	"During the discussion only the quality of an argument is important, not which person is bringing it forward."	+3.14	+2.38
2	"A respectful discussion atmosphere is important for me to present my ideas to the group."	+3.28	+3.08
3	"I experienced even during controversial parts of the discussion no hurtful or irrelevant arguments."	+3.3	+3.2
4	"I could present my standpoints to the others during the course of the [BEKO] events."	+2.65	+2.78
5	"Public participation programs like the BEKO are a good way to let the public participate in political decisions."	+3.16	+2.61
6	"In the future I would take part again in a public participation program like the BKEO."	+3.49	+2.94

while stakeholder organizations have deep-rooted, but contrary reasoning. Here the BEKO is fulfilling its purpose but at an inefficient cost-value ratio. Hence the rating.[4]

These results show that the moderation has a clear task to keep discussions factual and fair in situations where reservations against each other are prevalent. But successful participation like the BEKO can solve this and even set the foundations to a higher satisfaction with political decision-making and open up reservations for future participative formats.

8.8 Conclusion

8.8.1 BEKO

In total the citizen tables produced 272 recommendations, the stakeholder roundtables produced 334, and the reflection table of the spokespersons for each topic created additional 145 recommendations and comments for the ministry. As a result of this productive involvement process the administrators at the ministry took three months to prepare a report specifying which measures they will adopt, modify, or reject (and the reasons why). In the end of 2014 the State government passed the State Climate Protection Plan which was based on the recommendations of the stakeholders and the citizens.

The participants of all groups stated that they would continue to participate in participation procedures such as the BEKO on important political issues in the future. However, such procedures should be carried out purposefully and with reasonable effort. They also

[4] Written statement of Participant 436, Variable 245 s, translated but by the author.

expressed the expectation that more publicity work should be done in future to motivate more women, migrants, and young people to take part in the citizen tables.

The participants also expressed that sufficient time should be scheduled between sessions to avoid time constraints. In addition, efforts should be made to formulate the recommendations and instructions directly and collectively in the meetings and to make visible for everyone (e.g., with a projector). This would require less coordination effort.

In addition, the provided material (in this case the IEKK working draft) should contain information on the effects and costs of a measure, its affordability, the actors involved, and so on.

In order to strengthen the confidence of all participants in the seriousness of the participatory procedures, it was important for the participants of the BEKO to make the results of the examination of the recommendations comprehensible and, in the case of a negative result, to give—as far as possible—reasons for the rejection.

8.8.2 Evaluation

Evaluation of a multimodal participation program like the BEKO leads to the necessity of multimodal evaluation strategies, like written surveys together with guided interviews are the core data source accompanied by observation (including online parts) and media analysis. But such a complex evaluation strategy not only demands more resources, time, and effort. It offers through its multiple perspectives insight into the participation process and its converting power on participants as well as its initiators. The BEKO evaluation could prove that it

- transformed fear among different societal groups about each other into a positive, joint perspective on participation and democratic values.
- showed the administrative departments not only the high level of interest to participate but also the interest to learn about alternatives and necessary compromises in decision-making by the population. This shifted the picture away from the infamous "Wutbürger"[5] to one of citizen as being positively engaged in playing a part in political decision-making beyond voting.

8.8.3 Lessons learned

In addition to the direct results for the transformation of energy and climate policy in Baden-Württemberg, the BEKO experiences can be used to obtain transferable information for the design of future participation projects:

- *Complexity*: At the BEKO four groups were involved at the same time:
  - randomly selected citizens at four citizen tables on the topics of "power supply," "households" (2 ×), and "traffic,"
  - the self-initiated citizens (5th citizen table),
  - the general citizenship through online participation, and

[5] A German term formed in 2010 to describe citizen who are highly visible in protest against any decision made by the authorities and a strong demand for more "power to the people."

- the organized representations of associations, interest groups, and organizations involved in the seven stakeholder/association tables ("power supply," "private households," "industry," "traffic," "trade, commerce, service," "agriculture, forestry, and land use," and "public sector").

  Despite the high level of organizational effort associated with this wide spectrum of people, the sessions were successfully carried out. Furthermore a connection to and a bond between the groups was established.

- *Selection of the associations, interest groups, and organizations*: In the selection of the invited associations, interest groups, and organizations there were minor supplementary requests by organizations that were initially not invited. If the working capacity of the group was not unacceptable due to its expansion, the requests could be considered. If the participation of other organizers was not sensible or not possible for other reasons, a contact mediation was offered with interest-like groups (interest representation).
- *Participant selection (citizens)*: A random telephone call ensured that every resident of Baden-Württemberg had almost the same chance of participating in the BEKO procedure. However, this did not prevent that some sections of the population did not participate at the BEKO according to their proportion in society. Here it could be considered whether the voluntary principle (as an indicator for the willingness to participate) should be expanded by an element of targeted recruitment.
- *Procedural rules*: It has proven as useful to send the participants the "procedural rules" with the registration confirmation and to point out that these rules are the basis for the participation.
- *Learning effects*: The participants of the citizens and the stakeholder/association tables expressed a very positive feedback after the joint sessions, because the reflection sessions had been on an equal level and declared that they had learned from each other. They spoke of a fruitful and valuable exchange in which similarities could be identified, which were not known before. Even a joint representation of interests in future topics was considered. Learning effects were also found among the administrative staff who had participated in the BEKO as experts: They said that they had a certain skepticism about the expertise of citizens before the BEKO. This view changed through the participation in the BEKO in so far, as some of the administrative staff were pleasantly surprised by the competence of the citizens and therefore would welcome their participation in future proceedings.
- *Accompanying public relations*: In order to establish the transparency of a participatory process, it is important not only to create an informative homepage but also to communicate its existence it in a suitable and broad manner. Accompanying public relations activities are therefore important. So it was not surprising that many participating citizens said that they first learned about the BEKO through the telephone recruitment. Although all cities and municipalities in Baden-Württemberg were previously asked to publish information on the participation process in their local magazines, their limited effect must be noted.
- *Comprehensive documents*: In a participation process, it must be ensured that the discussed documents are formulated as comprehensibly as possible. Therefore communication expertise should be consulted and considered at an early stage.

- *Timely dispatch of the documents*: The documents, minutes of meetings, and so on should be sent in time. In the case of the BEKO an attempt was made to send the respective documents 1–2 weeks before the meetings, which turned out to be in part too brief.
- *Time buffer*: When designing the participation process, appropriate time buffers at important points should be installed. At the BEKO, for example, in one sector another meeting was required, which could be realized due to a generous overall time planning. In addition, it was important for the participants to have enough time between the sessions to be able to deal intensively with the results or to discuss the results with colleagues from their institutions.
- *Connectivity of the results*: At the BEKO the recommendations were examined by the responsible departments. In order to make the effort for the examination of the results affordable a clarification of the scope and form of the presentation of the recommendations (before the formulation of the recommendations) turned out to be expedient. At the BEKO therefore all recommendations were initiated, for example, with the phrase "the table recommends the state government, ..."
- *Announcement and feedback regarding the recommendations*: Important for the confidence in participation processes is a transparent, comprehensible, and pre-procedural representation, how the results and recommendations will be handled. This claim should be credibly fulfilled, for example, through a detailed documentation of the results of the review process of the recommendations. It should make clear which of the recommendations were taken into account and which were not. If a recommendation was not included, a comprehensible justification should be provided.

Reference

Ministry of the Environment, Climate Protection and the Energy Sector Baden-Württemberg, 2012. BEKO Information Brochure "We are breaking new ground! Join us!" <https://um.baden-wuerttemberg.de/fileadmin/redaktion/dateien/Altdaten/202/Anlage_IEKK_und_BEKO.pdf> (accessed 14.07.19.).

CHAPTER

9

Digital tools in stakeholder participation for the German Energy Transition. Can digital tools improve participation and its outcome?*

Anna Deckert[1], Fabian Dembski[2], Frank Ulmer[1], Michael Ruddat[3] and Uwe Wössner[2]

[1]Dialogik Non-profit Institute for Communication and cooperation research, Stuttgart, Germany [2]High Performance Computing Center, Stuttgart, Germany [3]Center for Interdisciplinary Risk and Innovation Studies (ZIRIUS), Stuttgart, Germany

OUTLINE

* The paper results of the authors' collaboration in the Reallabor Stadt:quartiere 4.0 ("Living Lab: City Districts 4.0").

The Role of Public Participation in Energy Transitions
DOI: https://doi.org/10.1016/B978-0-12-819515-4.00009-X

9.1 Introduction

Regarding the German energy transition (Energiewende), there has been consensus about its necessity and desirability for many years. While 93% of citizens surveyed by the German Renewable Energy Agency (AEE) support a stronger expansion of renewable energies in Germany, a smaller percentage (63%) supports the construction of energy plants in their neighborhood. Similarly, 78% regard grid expansion as important. Still, only 32% support it in their neighborhood, while 36% are undecided and 30% do not support it (AEE (Agentur für Erneuerbare Energien), 2018).

This phenomenon—also known as NIMBY (Not in my backyard)—often occurs in the context of infrastructure expansion (Menzl, 2014). To avoid or mediate conflicts, involvement of stakeholders[1] is often regarded as a promising approach, even though it cannot work wonders, either (Spieker, 2018: 77–80).

Next to setting a basis for acceptance, participation processes can activate and utilize capabilities of stakeholders for the planning and implementation of large-scale projects, such as the construction of energy transition infrastructure or smaller projects in urban development. The aim is to improve the quality of decisions that are ultimately made by political representatives (such as the municipal council) by involving the knowledge and preferences of a diverse group of people and institutions, but taking part in planning processes for renewable energy infrastructure is, of course, not the only way to participate in the German energy transition.

In a broader sense, this can happen in different ways, such as financial investments, putting photovoltaics on the rooftop or using energy more efficiently (Sonnberger and Ruddat, 2016). About half of the installed capacity of renewable energy in Germany is in

[1] In addition to interest groups, companies or scientists who are traditionally perceived as stakeholders, often citizens want to participate in the design of these projects at an early stage, instead of having to call for their participation in cases of conflict. In the following, the term "stakeholder participation" refers to the involvement of people and institutions who have a legitimate interest in the course or outcome of a project, during the preparation stage of a decision. Thus citizens are explicitly assigned to the group of stakeholders.

the hand of citizens (AEE (Agentur für Erneuerbare Energien), 2014). They play an important role in the transition process, in a broader and narrower sense of participation (Ruddat and Renn, 2012).

In this chapter, we concentrate on the narrower version of participation: Taking into consideration the increasing complexity and the chances of digital transformation, we address the question of the extent to which digital instruments can support the successful implementation of participation processes.

As stated by Spieker (2018: 85), regarding visualization technologies, digital tools can support the assessment of plans for the transition projects, but they need to be embedded in "serious dialogs" that are valued by political decision-makers to improve participation and the communication between decision-makers in politics, administration, and citizens. In the desired political process, representatives consider different arguments brought up. So, when proposals cannot be implemented due to content-related reasons, it is plausible, but if proposals are dropped without appropriate consideration or just because it seems politically opportune, participation is worthless, and there is no basis for the digital tools to be of any support. Therefore they cannot solve implementation problems that are resulting from governance structure. It is a prerequisite of successful participation that the administration and local councils show a willingness to consider and, when appropriate, implement recommendations.

Before discussing the contribution of digital tools to the success of participation processes, general success factors beyond this basic "willingness to implement" need to be set out. In the following, the chapter focuses on the potential of visualization tools. It describes and evaluates two case studies, the Forbach Pumped-storage Power Plant and Expansion and Urban and Traffic development in Herrenberg, from two angles: the perspective of the executing organization EnBW Energie Baden-Württemberg AG (a publicly traded electric utilities company) on the one hand and the involved citizens in the case of Herrenberg on the other hand.

9.2 Success factors of stakeholder participation

A perfectly planned stakeholder participation process, which satisfies all parties involved, does not exist, but from our experiences in other participation processes, we know that some factors do make success more likely. They are to be seen from a rather operational perspective (for a more theoretical view on success in terms of evaluation criteria, see e.g., Goldschmidt, 2014).

9.2.1 Clear mandate and shared understanding of the purpose of participation

One, if not the decisive factor for success, is a clear understanding on all sides of the objective or function of stakeholder participation. The objective may be to activate the knowledge of different actors, to qualitatively improve a project/measure, but it may also be the goal to ask for preferences, to get an opinion of different stakeholders (Alcántara et al., 2014).

If those responsible for a participation process have agreed on its function (maybe even in dialog between politics, administration, and citizens by applying the trialogic principle), it is important to provide the involved stakeholders with a clear mandate. Everyone involved must know at all times if they, for example, discuss or decide on the "whether" or on the "how" of a measure (Goldschmidt, 2014: 239). It is necessary to explain to all stakeholders what is going to happen with their recommendations and proposals and whether they can or cannot be implemented.

9.2.2 Compatibility of political and administration processes

Those in power, whether in government or in administration, should have a concept for the integration of other stakeholders' ideas into the respective project (Hebestreit, 2013: 78). One prerequisite for successful dialogs is the ability to connect newly created results to administration and politics. For this purpose, the preparation of proposals must be directly interlinked between politics and administration within the scope of a participation process (trialogic principle). Methods of agile administration, such as modern techniques of visualization and organization (e.g., Trello or Kanban), can be used to circulate results to the administration and local council.

9.2.3 Structure of participants

The selection of participants is fundamental for the success of a stakeholder participation process (Hebestreit, 2013: 80). It is important to ensure a balance between all interests. All groups, such as associations, should be given the opportunity to state their objectives and expectations. In doing so, their intention for constructive participation can already be identified (Ulmer and Sippel, 2016). A selection by a randomized process has the advantage that not only the "usual suspects" participate. It offers the chance to also recruit young people or people with a migration background, etc. (Ulmer and Sippel, 2016).

Depending on the subject of participation, it may also make sense to recruit not only randomly selected citizens but also, for example, mainly young people or people who are particularly confronted with the project in question in their everyday life.

9.2.4 Transparency and good moderation

It has repeatedly shown that transparency during a participatory procedure is a key component for success (Hebestreit, 2013: 83). Good planning and clear framework conditions must be complemented by transparent communication (Goldschmidt, 2014: 239–240). Informing participants and noninvolved public about the procedural steps and the interim results helps to prevent doubts about the participation and a possible failure. This requires a consistent moderation that keeps track of every step, establishes a structure, sets rules, and summarizes results (Goldschmidt, 2014: 195–196).

9.3 How can digital tools contribute to the success of participation processes?

Digital tools can support participation, but they need to be embedded in "serious dialogs" that are valued by political decision-makers (Spieker, 2018: 85). They can improve communication between decision-makers in politics and administration and citizens by reducing the complexity of technical knowledge, simplifying data processing, illustrating results in a more intuitive way, and making participation more attractive. Still, the results need to be acknowledged by the decision-makers.

9.3.1 Digital tools for offline participation

Talking about digital tools, we are not referring to voting on digital platforms. We are talking specifically about tools that support offline participation. The following categories of digital instruments can support the success factors of such offline participation:

- information platforms (websites, social media, etc.),
- motivational tools to stimulate discussion (films, videos, blogs, etc.),
- simulations and scenarios (e.g., for computer-aided and evidence-based urban design),
- Virtual Reality (VR; e.g., 3D experiences of building projects), and
- voting tools [tele-dialog system (TED), apps, etc.].

Such digital instruments have been tested and described in the author's collaborative project Reallabor Stadt:quartiere 4.0 (Living Lab: City Districts 4.0). The descriptions and learnings can be accessed via www.digitale-mitwirkung.de.

9.3.2 Improvement of decisions through digital tools

Digital tools in stakeholder participation can help in at least three ways to improve the quality of decisions eventually made by the representatives (see also Spieker, 2018: 84).

9.3.2.1 Make a contribution to diversification of participants

Digital communication tools can communicate the request for stakeholders to participate. Social media can advertise events and distribute information (Robra-Bissantz et al., 2017: 470), especially when young people are a target group, but these methods still exclude parts of the population.

Other digital tools, such as videos, VR, or voting tools, can facilitate the activation of younger generations and other groups who rarely take the opportunity to participate in the deliberation. The playful simulation approach provides an additional incentive to participate.

9.3.2.2 Reduce complexity to make plans easier to assess for laypeople

Digital tools can help to activate the knowledge of different stakeholders for the improvement of a project. According to the companion modeling approach (Étienne, 2013), visualization, as an example for digital tools, has three desired effects: creating knowledge, improving the interaction with others, and creating frame of reference for discussions between participants.

For instance, using a Digital Twin in VR (see Case Study below) makes a possible consensus among participants with different backgrounds more likely. This is furthermore connected to a common learning process linked to educational aspects. As Glaeser et al. (2006) rightly point out, better-educated citizens are more likely both to preserve and strengthen democracy. In general, visualization of complex processes and data for the participation of heterogeneous groups is essential, but the outlay to realize a 3D projection for large groups of participants and finally the realization of such big events is also a matter of budget. Furthermore, the hardware needed is yet not easily accessible, but a Digital Twin offers great potential in the field of digital tools. Loaded with quantitative and qualitative empirical data, it is a promising approach to manage the complexity of such projects and to involve citizens in the planning process.

The term "Digital Twin" has been coined and used first in relation to mechanical engineering where they are mainly applied for several years. Digital Twins are digital representations of material or immaterial objects, such as machines from the real world. They enable comprehensive data exchange and can contain models, simulations, and algorithms describing their counterpart and its features and behavior in the real world (Kuhn, 2018). A Digital Twin can be best characterized as a container for models, data, and simulation.

One example could be a Digital Twin of a renewable energy facility visualized in VR. Stakeholders can contribute their ideas directly and also receive immediate feedback regarding feasibility and impacts. This is also important in the sense of transparency, since a faster and more direct feedback is possible. Fiukowski et al. (2017) examined the ability of digital tools to empower stakeholders in participation processes throughout five stages of participation (information, consultation, cooperation, delegation, and self-sufficiency). They conclude that the assessed tools "helped them [the interviewees] to better understand the field and to feel comfortable participating in discussions with subject-matter experts" (Fiukowski et al., 2017: 7). Participants without prior expert knowledge thus had the chance to take part in the deliberation.

9.3.2.3 Improving the presentation of results make them easier to interpret for decision-makers

Borras and Edler (2014: 28) state, that "[t]he complex nature of socio-technical systems makes the participatory and effective governance of change in socio-technical systems more dependent on knowledge of citizens and experts alike." Visualization of results of stakeholder involvement can help decision-makers to understand and assess the contributions of experts and citizens. Especially in case of complex coherencies, for example, of a local energy system, decision-makers in politics and administration can benefit from the explaining quality of digital tools. The easier and better understanding of scenarios and workshop results improves their political connectivity (Fiukowski et al., 2017: 7).

9.3.3 Digital tools and participation culture

Some digital tools can help to lift the spirits, make the process of stakeholder participation as enjoyable as possible for the participants, and thus help to establish a culture of participation (Robra-Bissantz et al., 2017: 471). Videos can inform stakeholders about certain issues in an appealing manner. If voting is part of the process or a quick impression of opinions is regarded helpful, a so-called TED can lead to quick voting results. Hence, the tools support offline participation. Only when people come together, a shared culture of participation and a shared creative spirit can emerge. Mere online participation runs the risk of being meaningless. In addition, in conflict situations (e.g., the search for a location of a wind power plant), directly affected opponents are mostly over-represented in online processes. Those whose individual interests are most vulnerable to the project tend to invest the most energy to mobilize people to participate in the online process (e.g., via social media). In the case of offline processes, it is easier to ensure that different interests and perspectives are included to achieve a diverse culture of participation.

9.4 Digital Twin of the Forbach Pumped-storage Power Plant

For the Black Forest region in Baden-Wuerttemberg, we created a Digital Twin in VR in cooperation with the energy supply company EnBW Energie Baden-Wuerttemberg AG for collaborative design and civic engagement in the context of the German energy transition—targeting the pumped-storage power plant in Forbach and the expansion of the historical Rudolf-Fettweis plant. EnBW aims to renew and extend the 100-year-old existing power plant to fully use its current potential by increasing the efficiency of existing hydropower stations and by providing the storage of regenerative produced energy (EnBW, 2019).

Energy transition is of national interest and essential for reducing emissions to reach climate goals like described in the Erneuerbare-Energien-Gesetz (Renewable-Energy-Law) (2000) or the Framework Convention on Climate Change (United Nations, 2015). Storing energy on demand might be one solution to achieve the German energy transition: on a long-term basis, pumped-storage power plants can store large quantities of energy and feed it back into the grid at short notice if required (EnBW, 2019).

The pump-storage power plant is planned to serve as power storage with a diurnal cycle: if surplus energy from electricity networks is available and the energy production higher than the current need, water is pumped into the basins and by this transformed to potential energy. The total efficiency of the pump-storage in Forbach is approximately 75% (EnBW, 2019). In particular, production from wind turbines and solar power plants is highly dependent on environment and climate conditions and cannot be steered. When the share of renewables increases, storage demand also increases. On the other hand, projects of this size have an enormous impact on local and regional development, quality of life, and the environment.

The initial situation was very complex, as the planned large-scale project is situated not only in a historic city but also on the boarder of the Black Forest national park. In addition, the historical power plant "Rudolf-Fettweis-Werk" is under monument protection. This was the reason why the approach focussed on a regional planning procedure (Raumordnungsverfahren), which EnBW implemented voluntarily. Their objective was to involve local governments, public authorities, NGOs, and citizens already at an early stage of the conception phase, before the approval procedure and during the whole planning process. For this purpose, the visualization department at High Performance Computing Center Stuttgart (HLRS) accompanied this process by developing a Digital

Twin visualized in VR environments. This approach is meant to support stakeholders in understanding complex relationships and in communicating their interests. The knowledge of all involved parties is regarded much more accessible to the planners and can be used to improve the planning.

9.4.1 The Digital Twin

The implementation of Digital Twins for towns only recently has been discussed (Batty, 2018). We developed a Digital Twin, which can be used across all scales, on multiple layers and in different categories of data, in virtual and augmented realities, for collaborative and participation processes, focusing on planning and decision support (Fig. 9.1 and 9.2). For all participation processes, we used collaborative VR environments. These environments,

FIGURE 9.1 The Digital Twin of Forbach: traffic and construction noise simulation, technical equipment, existing power plants infrastructure and building, built environment/Forbach, geological model, and underground power station (Dembski et al., 2019).

FIGURE 9.2 Public participation in Forbach using a mobile collaborative VR environment (Dembski et al., 2019).

such as stereoscopic back projection units, large 3D displays respectively tiled display systems or Cave Automatic Virtual Environment (CAVE), help different participants with diverging professional and personal backgrounds to use the system simultaneously. Discussions can be enhanced, and joint consensus for solutions can be approached (Dembski et al., 2019).

The implementation of Digital Twins in VR helps citizens to understand it better and enables them to participate. In general, 3D models and visualizations are used to overcome the lack of communication and therefore support transparency and decision-making.

9.4.2 Methods and data

For this project, the Digital Twin is set up as follows: (1) a morphological model including a digital terrain model, aerial photographs, and the built environment in form of a 3D city model (LOD1); (2) 3D models of the existing power plants infrastructure and building information models (BIM) of the planned interventions; (3) a geological model including boreholes and rock strata analysis; (4) 3D models of the technical equipment (such as turbines, pumps, valves, and fish lifts); (5) mechanical simulations of the water flow through turbines and inside the water storage system; and (6) environmental impact simulations, such as construction and traffic noise simulation.

The Forbach Digital Twin is a strongly process-oriented project. It started in 2011 and has been continuously developed and improved ever since. Data provided are regularly updated, and information layers are added in alignment with the planning progress and requirements with regard to construction phases, concomitant participation processes, implementation, and evaluation. The construction and maintenance of the Digital Twin is a long-term process, having eventually an open end.

For the Digital Twin, we used the Collaborative Visualization and Simulation Environment (COVISE). It is an extendable distributed software environment to integrate simulations, postprocessing, and visualization functionalities in a seamless manner.

9.4.3 Virtual reality environments support the collaborative and participative processes in Forbach

With the use of VR environments, numerous participants were involved in collaborative and participative processes since 2011. Both top-down and bottom-up approaches were pursued: the EnBW engineers used the Digital Twin in the CAVE at the HLRS first for collaborative planning processes and technical meetings together with specialist planners and engineers to gain an overview about the content-related and spatial complexity of the project. VR helped the participants to gain a better understanding about functions and implementations during the different planning phases and to help solving technical issues (Gommel and Wössner, 2019). At a later stage, the EnBW engineers used the Digital Twin to inform the management level during internal workshops, making their decisions transparent on the company level but also for supporting financing processes and legal procedures. 3D visualization in VR turned out to be eminently suitable for conveying complex contents to experts from different disciplines and to improve imagination.

During the entire planning stage, many groups of stakeholders were involved in participation processes under the lead of the EnBW project management team, such as the BUND (German Federation for the Environment and Nature Conservation), the local visitor center, Schwarzwaldverein (black forest association), Industrie- und Handelskammer (Chamber of Industry and Commerce), local governments, and various NGOs. Topics proposed by the participants were, for example, concerns about construction noise pollution and traffic during the implementation, visibility of the power plants constructions and buildings and other mainly environmental topics. By means of visualizations in VR environments, it was possible for the participants to gain an overview and insight by taking specific spatial positions and identifying the needed information levels.

In the framework of "Reallabor Stadt:Quartiere 4.0," also large-scale public participation processes were conducted. In October 2011, 400 participants visited a large-scale presentation of the Digital Twin, similar to a 3D cinema but in an interactive way. During the event, the citizens could ask questions and receive customized answers by navigating to an appropriate perspective or information level, for example, visibility concerns or construction noise.

A second large-scale public participation process was performed as part of the 100-year anniversary of the historic power plant in 2018: more than 1300 participants used a mobile collaborative VR environment consistent of a 3D projector, a back-projection screen and associated hard- and software.

9.4.4 The executing organization's view on the "Digital Twin"

The lead manager of the Forbach project, Ulrich Gommel, emphasized that this innovative way of using Digital Twins and VR for public participation processes helped a lot: to inform parties concerned, to increase interaction, to identify the potential for conflict at an early stage, to respond to concerns, to make immediate corrections, and to gain a clearer understanding of potential problems on the part of all participants. Even if parties are fundamentally opposed to a project, these participative processes in VR could significantly contribute to objectivize the discussion by informing and forming a platform for the rational exchange of arguments (Gommel and Wössner, 2019).

For the creation of a Digital Twin, BIM and many other data in 3D are a basic requirement. People responsible at EnBW aim to implement these methods in their rule-set on architectural competition and as a standard for specialist (ibid).

Visualization in VR can support different processes in the framework of the German energy transition. It is suitable to provide vistas from various specific locations of choice to provide information and involvement already at an early stage of the planning phase of wind farms, hydropower plants, solar power plants, and associated infrastructure like power lines. According to Gommel and Wössner (2019), 3D models in VR are, in sum, an excellent way to illustrate complex planning projects and content to laypeople. It also supports experts in collaborative planning processes.

9.4.5 Further development of virtual reality environments

Another interesting use could be related to educational aspects and in strengthening the democratic decision-making ability: in the Digital Twin, simulations could provide information about the energy demand in the region and the potential of renewable energy for the regional supply in the catchment area. Another focus could be set on the virtual comparison of different energy sources, such as hydropower versus coal-fired power plants, their environmental impact, and the countrywide distribution of electrical energy to support the understanding of the German energy transition with regard to teaching and youth participation. At a technical level, also sensor network data providing environmental and climate data and data from Volunteered Geographic Information could be applied (Dembski et al., 2019).

By its nature of a model, and therefore an abstraction of reality, a Digital Twin does not include all real-life information. The objective is to achieve similarities to the real world and a level of detail accurate enough for concrete (but complex) problems. Furthermore, Batty (2018) states the strong need for additional social, economic, and environmental data.

Whereas there is scope for improving these Digital Twins by implementing urban, regional, and economic analyses, traffic simulation and visualizations of the energy demand, and flow in the region, furthermore, there is a need to simulate and illustrate social aspects with the help such as synthetic population models.

We are continuing our research in the context of Global System Science related to these areas and purpose. Therefore we are currently working on an integrative toolbox for global systems analysis. The integration will be cantered on recent methodological advances in the construction and use of synthetic populations. These synthetic populations provide models of given populations, typically of humans, but amongst others also of cars and buildings. A synthetic population is based on individuals that are different from the actual ones, but in a way that the population matches the empirical one in the distribution of attributes and relations that matter for the problem at hand. In our current research, we aim to provide generators for such populations at a worldwide scale but also at smaller scales. The Digital Twin of Forbach forms just one piece in the big puzzle.

9.5 The citizens' view on "Digital Twins" in the case of Herrenberg

User surveys proved the usefulness of this kind of model for citizens also in the case study of Herrenberg. Herrenberg is a medium-sized town in the southern Germany with a population of approximately 30,000 people. A main road crosses the medieval core at a point where the traffic of three highways merges. Thus this area is exposed to high traffic volumes and to environmental pollution by emissions. To solve this urban challenge, the city of Herrenberg is currently developing an integrated mobility plan (IMEP 2030) to serve as the guideline for the mobility development over the next 15 years. Public participation processes were implemented by "Reallabor Stadt:Quartiere 4.0" to support this and other urban development processes. For this purpose, we created a Digital Twin and visualizations in VR using different devices, such as back projections generated by 3D projectors for the visualizations, 3D glasses (active shutter 3D system) and a tracking system for navigation and proper representation of perspective for the viewers during various events on site (Fig. 9.3).

9.5.1 Method of evaluation

The Digital Twin of Herrenberg was also presented during the event "Morgenstadt-Werkstatt meets Digitalakademie@bw" at Fraunhofer Institute for Industrial Engineering (IAO) in Stuttgart in December 2018 and on the New Year's reception in Herrenberg in January 2019. Visitors could interact with the model in VR environments using 3D glasses and explore the different parts of the simulation, such as fine dust, wind flow, Space Syntax Analysis for traffic planning, movement patterns of bikers and walkers,

FIGURE 9.3 The Digital Twin of Herrenberg: civic science data, space syntax analysis, movement patterns and urban morphology (Dembski et al., 2019).

and danger zones for bikers. Groups of roughly 10 people used the VR environment for about 10–15 minutes each.

After experiencing the virtual model, the visitors filled in short questionnaires. Topics were the perception and evaluation of different aspects of the model, expected positive effects for urban planning and public participation processes, former experiences with public participation processes, and several open questions about, for example, the benefit of the virtual model or missing elements in the model. The questionnaire can be found in the annex. Almost 40 people participated (10 in Stuttgart and 29 in Herrenberg).

9.5.2 Results

The participants highly approved the virtual model: the bold line in Fig. 9.4 located on the far right positive end of the scale indicates this. They judged the model to be very comprehensible (+2.37; on a scale ranging from −3.00 to +3.00) as well as interesting (+2.54). Furthermore, they rated concreteness very highly (+2.14). Moreover, the model appears to be simple (+1.56), clearly arranged (+1.92), and entertaining (+1.89). Although these three categories score lower than the other three, they are clearly in the positive area of the scale.

The overwhelming majority of the participants stated positive effects of virtual urban models for local planning as well as for public participation processes. 95% believed in the usefulness for local planning, 97% said it would be good for public participation. One reason for this remarkable result could be the professional background, since at least the event "Morgenstadt-Werkstatt meets Digitalakademie@bw" at Fraunhofer Institute for Industrial Engineering (IAO) in Stuttgart was very attractive to people interested in innovations. And indeed, many people working in the engineering field, in information

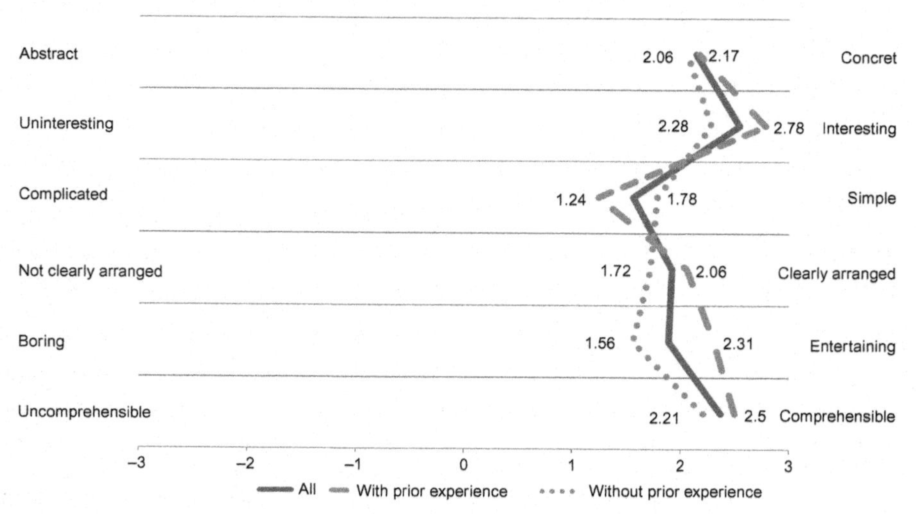

FIGURE 9.4 Perception and evaluation of the virtual model of Herrenberg.

technologies or the design area, attended the events in Stuttgart and Herrenberg, but also, teachers, policemen, or a nurse attended, so all in all a quite broad range of professions.

We also asked for former experiences with public participation processes. A total of 46% have already attended participation events as citizens and 13% as official local representatives. Three people said that they participated as both citizens and official local representatives. Therefore half of the interviewed people are principally able to judge the usefulness of virtual models on the background of their experience with past participation processes.

This leads us to the question if people with prior participation differ in their evaluations of virtual models from people that have not yet participated. We suspect that virtual models are up to now not very common in typical participation formats (this was one reason for our own research project Reallabor Stadtquartiere 4.0). 2D plans of city areas and planning concepts in mainly technical language still dominate the scene. This means, people with experience in these classical forms of participation can compare them im- or explicitly to our new and innovative virtual model. Our analysis shows that the group with former participation experience judged the virtual model on nearly every dimension on average more positively compared with the group without such experience. The first group is represented by the broken line in Fig. 9.4, and the latter is represented by the dotted line. The only exception is the dimension complicated—simple. People with former experiences judge the model to be more complicated on average compared with people without such experiences. All in all, VR is evaluated better within the group with prior experiences in participation processes.

The participants stated in the open format of our survey a great potential for visualizing situations, consequences, planning processes, etc. According to them, concreteness and transparency are two main advantages of virtual models that can be very important in interaction processes between official local representatives and citizens. This is especially true with respect to the simple and understandable communication of complex and long-term processes in urban development.

Citations:

Situations/circumstances can be presented in many perspectives.
Better imagination of consequences/implications.
Complex planning processes can become more concrete.
One can better imagine the spatial impact.
Simple presentation, everyone can imagine the plans better.

The research project Reallabor Stadtquartiere 4.0 (Living Lab: City Districts 4.0) addressed this by investigating the potential of digital participation methods and visualization tools for involving citizens in the planning process of city districts. If the perception and evaluation of the workshop participants in Herrenberg and Stuttgart is true for other social groups in society or society (which still needs scientific investigation), it could be successfully used in participation processes.

Relatively complex information about the respective project could be transferred from official local representatives and other planning experts to the citizens using virtual models. The citizens could then generate ideas fitting almost exactly to the proposals. The virtual model could also be an answer to one of our main research questions in the "Living Lab: City Districts 4.0" concerning new tools for interaction in planning processes (Which tools can help to improve the connection between the generation of ideas and urban planning as well as the interaction between citizens and administration?). It is important to mention that this is only the perception of the participants. We would need a scientific test under real-life conditions with the comparison of analogs and digital instruments of information and participation to come to more valid conclusions.

Missing elements of the model mainly had a connection to traffic. Participants wanted to see more about traffic flows, parking situation from Monday to Saturday or the areas in town used only by walkers and bikers. Another point was the accessibility of the model. It should be possible for citizens to use it either by visiting a public place (like a townhall) or in their homes using 3D glasses. Potential for further improvement was stated with respect to the quality of the graphical presentation. It should be more realistic, like in Google Earth. On the other hand, we had very enthusiastic comments on the technical possibilities of the model.

9.6 Conclusion

A clear understanding of the stakeholder's role and the mandate of those involved are important success factors of stakeholder participation. The connectable and transparent design of the process is crucial to its success. With regard to the integration of various interests, attention must be paid to a diverse participant structure.

Digital tools can support the creation and maintenance of a culture of participation. The same counts for the success of stakeholder participation processes by contributing to the diversification of participants, a reduction of complexity for laypeople as well as for politicians and administration and a simplification of the interpretation of results. Personal contact remains essential for successful participation procedures and should not be replaced by digital tools, but should be supported. Participation processes for energy transition projects or other complex transition processes that require technical knowledge can benefit from visualization tools like virtual realities.

The lead manager of the Forbach project appreciated the use of Digital Twins and their representation in VR for public participation processes. The tool was very helpful for informing all parties concerned and for increasing interaction. It also supported the identification of the potential for conflict at an early stage, the response to concerns, the gathering of immediate corrections, and the gaining of a clearer understanding of potential problems on the part of all parties involved.

Civic participants with experience in participation processes see added values when it comes to VR: they stated a great potential for visualizing situations, consequences, or planning processes in a very concrete and transparent manner. According to surveyed participants, concreteness and transparency are two main advantages of virtual models. Relatively complex information about the project can be transferred from official local representatives and other planning experts to the citizens using virtual models, such as VR. On this basis, they can generate ideas fitting almost exactly to the proposals.

Apparently, digital tools have quite some advantages for participation, but they cannot solve implementation problems resulting from governance structure. So, it is a prerequisite for successful participation that the administration and local councils are willing to consider and, where appropriate, implement recommendations.

References

AEE (Agentur für Erneuerbare Energien), 2014. Großteil der Erneuerbaren Energien kommt aus Bürgerhand. In *Renews Kompakt*. <https://www.unendlich-viel-energie.de/media/file/284.AEE_RenewsKompakt_Buergerenergie.pdf> (accessed 05.08.19.).

AEE (Agentur für Erneuerbare Energien), 2018. Klares Bekenntnis der deutschen Bevölkerung zu Erneuerbaren Energien. <https://www.unendlich-viel-energie.de/klares-bekenntnis-der-deutschen-bevoelkerung-zu-erneuerbaren> (accessed 05.08.19.).

Alcántara, S., et al., 2014. DELIKAT—Fachdialoge Deliberative Demokratie: Analyse Partizipativer Verfahren für den Transformationsprozess. Umweltbundesamt.

Batty, M., 2018. Digital Twins. Environ. Plan. B: Urban Analytics and City Sci. 45 (5), 817–820.

Borras, S., Edler, J., 2014. The Governance of Socio-Technical Systems: Explaining Change. Edward Elgar Publishing.

Dembski, F., Wössner, U., Yamu, C., 2019. Digital twin, virtual reality and space syntax: civic engagement and decision support for smart, sustainable cities. In: Space Syntax Proceedings of the 12th Space Syntax Symposium (316), Beijing.

EnBW, 2019. Pumpspeicherwerk Forbach—Neue Unterstufe, Antragsunterlagen zum Plansfeststellungsverfahren—Antragsteil A.V. Erläuterungsbericht, Stuttgart.

Étienne, M., 2013. Companion Modelling: A Participatory Approach to Support Sustainable Development. Éditions Quae.

Fiukowski, et al., 2017. Stakeholder Empowerment in Participation Processes of the Energy Transition—An Evaluation of Impacts of Simulation Tools. <https://reiner-lemoine-institut.de/stakeholder-empowerment-in-participatory-processes-of-the-energy-transition-an-evaluation-of-impacts-of-simulation-tools-fiukowski-et-al-2017/> (accessed 05.08.19.).

Glaeser, E.L., Ponzetto, G., Shleifer, A., 2006. Why Does Democracy Need Education? NBER Working Paper Series. National Bureau of Economic Research, Cambridge.

Goldschmidt, R., 2014. Kriterien zur Evaluation von Dialog- und Beteiligungsverfahren: Konzeptuelle Ausarbeitung eines integrativen Systems aus sechs Metakriterien. Springer Fachmedien Wiesbaden.

Gommel, U., Wössner, U., 2019. Telephone Interview on the Forbach Project, 22 March 2019, Stuttgart/Forbach.

Hebestreit, R., 2013. Partizipation in der Wissensgesellschaft: Funktion und Bedeutung diskursiver Beteiligungsverfahren. Springer Fachmedien Wiesbaden.

Kuhn, T., 2018. Digitaler Zwilling. *Informatik Spektrum* 40_5_2017. Springer-Verlag Berlin Heidelberg, pp. 440–444.

Menzl, M., 2014. Nimby-Proteste—Ausdruck neu erwachten Partizipationsinteresses oder eines zerfallenden Gemeinwesens? Stadt und Soziale Bewegungen 65–81.

Robra-Bissantz, et al., 2017. Das "e-" in Partizipation: Wie Digitalisierung und Vernetzung eine erfolgversprechende Partizipation ermöglichen. Springer Fachmedien Wiesbaden.

Ruddat, M., Renn, O., 2012. Wie die Energiewende in Baden-Württemberg gelingen kann. et.—Energiewirtschaftliche Tagesfragen 11, 59–62.

Sonnberger, M., Ruddat, M., 2016. Die gesellschaftliche Wahrnehmung der Energiewende—Ergebnisse einer deutschlandweiten Repräsentativbefragung. Stuttgarter Beiträge zur Risiko- und Nachhaltigkeitsforschung, Nr. 34, September 2016, Stuttgart.

Spieker, A., 2018. Stakeholder dialogues and virtual reality for the german energiewende. J. Dispute Resol. Available from: https://scholarship.law.missouri.edu/jdr/vol2018/iss1/9 (accessed 05.08.19.).

Ulmer, F., Sippel, T., 2016. Frühzeitige Öffentlichkeitsbeteiligung am Beispiel eines geplanten Neubaus einer 110-kV Hochspannungsleitung—Herausforderungen und Erfolgsfaktoren bei komplexen Beteiligungsprojekten, eNewsletter Netzwerk Bürgerbeteiligung.

United Nations, 2015. Framework Convention on Climate Change, Adoption of the Paris Agreement, Proposal by the President, Draft decision_/CP.21, Conference of the Parties, Twenty-first session, Paris.

Appendix

Questionnaire (translated from German)

1. The virtual model of Herrenberg is ...

	-3	-2	-1	0	1	2	3	
incomprehensible								comprehensible
boring								entertaining
not clearly arranged								clearly arranged
complicated								simple
uninteresting								interesting
abstract								concret

2. Do you think that virtual urban models can have a positive effect on public participation? ☐ Yes ☐ No
3. Do you think that virtual urban models can have a positive effect on local planning processes? ☐ Yes ☐ No
4. Did you ever attend a public participation process as a citizen? ☐ Yes ☐ No
5. Did you ever attend a public participation process as an official local representative? ☐ Yes ☐ No

PLEASE TURN OVER

6. What do you think are the benefits of virtual urban models in the context of urban planning?

8. What is your professional background?

7. What kind of information (for example specific data) is missing in the presented virtual urban model from your point of view?

9. Further notes and suggestions regarding virtual urban models:

CHAPTER

10

Citizen participation for wind energy: experiences from Germany and beyond

Gundula Hübner

MSH Medical School, Hamburg & Institut für Psychologie, Martin-Luther-Universität Halle-Wittenberg, Halle (Saale), Germany

OUTLINE

10.1 Introduction

Worldwide, in the industrialized countries, large-scale wind turbines and high-voltage power lines meet controversial discussions and resistance at the local level. Likewise, comparable acceptance factors and solution approaches can be observed, such as participative approaches, compensations, or technical device to mitigate emissions. Which participative approach can be applied depends on the different national legal frameworks and cultural norms, too (e.g., Geißler et al., 2013). As the details of national laws and regulations are too manifold and beyond the scope of this chapter, the present contribution focuses on the general impact of citizen participation on the local acceptance of wind turbines. Additionally, as

The Role of Public Participation in Energy Transitions
DOI: https://doi.org/10.1016/B978-0-12-819515-4.00010-6

cultural norms are diverse and nuanced even at local levels, they cannot be captured completely herein.

Recently, an increase in opposition is described in the media and even scientific publications while empirical knowledge on the relation of supporters, opponents, or indifferent local residents often remains vague. Consequently, the number of opponents might be overestimated and opposition perceived as the existing social norm. However, even at the local level, a majority seems to accept wind turbine projects—prior as well after construction (e.g., Firestone et al., 2018). Several factors influence the local acceptance of projects, such as perceived impacts on the landscape, birds and bats, property values or health (e.g., Ellis and Ferraro, 2016; Rand and Hoen, 2017). Even more, wind turbine deployment requires a cooperation among numerous stakeholders besides residents and land owners the legal authorities in charge of the permitting processes, business partner, nongovernmental organizations, or policy-makers. Indeed, a key factor for local acceptance of wind energy is the planning process, including the degree of perceived justice and fairness (e.g., Hoen et al., 2019). However, social acceptance is connected to a complex set of influences between individual stakeholders as well as communities, profit and nonprofit organizations, regulatory regimes, just to name a few (Ellis and Ferraro, 2016). The complexity of local acceptance makes obvious that citizen participation is fundamental but not the exclusive golden key to foster the development of wind energy.

Furthermore, in most countries, participation is associated with the planning process and the resulting acceptance of local projects but less with the impact of how the operating wind turbines are experienced (e.g., Hübner et al., 2019a; Mills et al., 2019). Since complains of residents about annoying wind turbine emissions cause negative information, uncertainty as well as opposition, the link between participation in the planning process and the later experience of the operating turbines needs to be highlighted as it may have a relevant impact on the public as well as on the local acceptance.

Before delving more deeply into the above addressed topics, the concepts of local opposition, acceptance and acceptability needs to be defined: Opposition can be understood as negative reactions and negative attitudes towards a local project while acceptance refers to the neutral or positive attitudes towards a local project and positive reactions. Acceptability indicates that a local wind project is not desired (negative attitude) but tolerated for diverse reasons, for example they consider the process by which it was approved to be legitimate (e.g., Batel et al., 2013).

10.2 Supporters—the reserved majority

In Germany as well as in many other countries in Europe (e.g., Austria, Denmark, Switzerland, and UK) and around the world (e.g., Australia, Canada, and the US), opinion polls consistently show that wind energy has broad public support. Likewise, positive attitudes toward wind power development are commonly found even toward specific wind power projects (e.g., Ellis and Ferraro, 2016)— with some exceptions, for example, in Denmark (Albizu et al., 2018). Despite these supportive attitudes, strong local opposition is observed, as reported in the media as well as experienced by wind project developers and local residents, indeed. However, the strength of oppositions does not reflect the local acceptance as typically, researchers have found local attitudes to be largely positive (e.g., Baxter et al., 2013; Wilson and Dyke, 2015; Fergen and Jacquet, 2016; Firestone et al., 2018; Hoen et al., 2019; Hübner et al., 2019a). These studies interviewed community members after local wind projects had been developed and assessed attitudes toward the development before and after the wind farm construction. The retrospective approach can be seen as a weakness but so far researchers have not had the chance to conduct an empirical survey before a wind development was annoyance or parallel to the announcement. Of course, opposition to or support of a proposed project, might be different than an attitude toward an existing project, although the correlates with opposition and support may be similar to those for negative and positive attitudes (Hoen et al., 2019). Additionally, most of the studies have not distinguished among residents who (1) moved in prior to construction and continue to live nearby, (2) moved to another community, or (3) arrived after construction. However, even when carefully distinguished between residents who lived near a project prior to its construction from those who move in after construction has commenced, the result pattern remains stable—the local majority seems to be neutral or in favor of local wind projects. For example, a recent nationally representative wind power perceptions survey of individuals living within 8 km of over 600 projects in the United States found most residents to have either neutral (41%) or positive (46%) attitudes prior to construction while about 7% were negative; this relation remained post construction (Fig. 10.1; Firestone et al., 2018). Comparable results were observed for Germany and Switzerland, for example (Hübner et al., 2019a). Additionally, the later studies observed that attitudes toward local projects were correlated with attitudes toward wind energy in general, reinforcing once again that "not-in-my-backyard" (NIMBY) might not be an appropriate explanation for negative attitudes locally (see Batel et al., 2015).

These results suggest that levels of community opposition are over- and acceptance understated. And indeed, opponents draw more to their position as they get more active. For example, in the representative US study given that almost six times as many individuals reported having positive or very positive attitudes compared to negative or very negative, a given opponent was slightly more than three times as likely to attend a meeting as

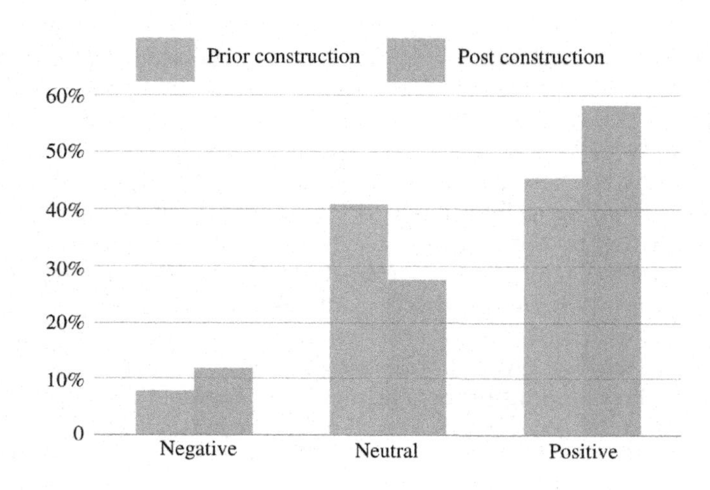

FIGURE 10.1 High levels of acceptance on local level—example from the US National Survey, presenting prior and post construction attitudes (Firestone et al., 2018).

a given supporter. Moreover, despite being outnumbered at meetings, opponents were more likely to speak (Firestone et al., 2018). In Switzerland, opponents participated more frequently in citizens' decision compared with other groups (Hübner and Löffler, 2013). And a recent study in Germany, in which residents of three different wind farm locations were included, provided comparable relations for local opponents to become more active compared with others (Hübner et al., 2019b). It seems that not only the opponents' activities influence the perceived social norm in the media and public but even local residents tend to underestimate their neighbors' positive attitudes toward a certain wind project (Hübner et al., 2019b).

It becomes obvious that innovative processes and tools are needed to improve participation in Germany and elsewhere, to activate neutral and positive community members, include a wider range of citizen and lead to more democratic legitimized outcomes. Throughout many countries, best practice examples and guidelines on how to design and lead participative process for wind energy and other renewable energies exists (see Ellis and Ferraro, 2016). However, citizen participation has to meet the expectations of the locals who will live closer to the wind turbines than anybody else. If the process is perceived as unfair or unjust, the acceptance will fade. This has potentially wide-ranging implications for the ownership and administrative regulation of wind energy. Despite the need for citizen participation, it should be kept in mind that local participation is not the solution to all problems. The concerns of visual intrusion, noise, and other environmental impacts have to be met as well as political concepts for a reliable energy transitions have to be provided to make wind projects acceptable at the local level.

10.3 The planning process and further factors influencing local acceptance

There is a wealth of literature on the factors influencing attitudes toward local wind energy projects (e.g., Wiersma and Devine-Wright, 2014; Cronin et al., 2015; Ellis and Ferraro, 2016; Firestone et al., 2018; Hoen et al., 2019). Most of this research is based on case studies, trying to understand public controversy (e.g., Jami and Walsh, 2016; Groth and Vogt, 2014; Firestone and Kempton, 2007), highlight best practices (Aitken et al., 2016) or place-based attributes (e.g., Pohl et al., 2018). Case studies focused on one or few specific wind projects that run the danger of case-selection bias and might capture case-specific rather than the factors that apply to typical projects (Firestone et al., 2018; Hoen et al., 2019). However, we do find a constant result pattern across studies and countries for the planning process fairness and openness to be related to local acceptance of wind projects. Furthermore, this finding is verified by the recent first-ever nationally representative analysis of this topic, conducted in the United States (Firestone et al., 2018; Hoen et al., 2019). Additionally, this US survey allowed a direct comparison with a larger European sample, including data from residents of several wind projects throughout Switzerland and Germany (Hübner et al., 2019a). Overall, following the cited retrospective studies besides the perceived degree of the planning-process fairness and justice, the landscape impact seems to be related closest to the acceptance of local wind projects.

10.3.1 Participation and compensation

A rich diversity of citizen participation methods exists to interact with communities and local residents, ranging from awareness raising over consulting to collaboration up to community empowerment and citizen-owned energy projects. By now it is well known that the degree of perceived fairness, distributive, and procedural justice is an important correlate with local acceptance of wind projects worldwide. Additionally, proper engagement can make project acceptable for residents who consider the siting process to have been fair although not all objections might have been solved. (e.g., Gross, 2007; Haggett, 2008; Walker et al., 2014a; Batel et al., 2015; Aitken et al., 2016; Firestone et al., 2018). Even more, residents who perceived the decision-making process as fair are less likely annoyed by operating wind turbines (Pohl et al., 2012; Pohl et al., 2018; Hübner et al., 2019a).

Besides the local acceptance perspective, the inclusion of citizens is fundamental due to their specific knowledge of their region and competences. One striking example for this is provide from a high voltage line planning process. In this process, the final route suggestions incorporated into the regional planning process would not have been considered without a participatory process (Kamlage et al., 2018). Opposite to this finding, in the case of wind energy, an emphasis on largely developer-led and controlled participative processes seems to exist at least in several European countries (e.g., Aitken et al., 2016; Cronin et al, 2015). Consequently, some researchers point to the problems which might arise from an unbalanced focus on commercial wind industry and call for larger changes in the planning regulations and policy regime to enlarge citizen participation (Ellis and Ferraro, 2016). This position can be described by the special case of Denmark, which was the entrepreneur of wind energy in the 1990s, but today meets strong barriers to wind project implementation. It is very briefly summarized that in Denmark, in the beginning wind turbines were owned by local citizens, farmers, and cooperatives. Investments were restricted to local residents living within a certain distance to the wind turbine, and amounts of shares were limited. In the late 1990s, these restrictions were removed. In the further development, with larger wind turbines, an increasing commercial wind turbine industry and political changes—for example the introduction of new planning rules and a tendering scheme—the possibilities for citizen or community ownership decreased drastically. As the local ownership went down, the local commitment to wind energy did so, too. It is concluded that local acceptance is reached only by community ownership models that support the local development (for details, see Albizu et al., 2018).

Findings from several countries underline the importance of community engagement and financial benefit distribution, for example, in Canada (Fast et al., 2016), France, and Germany (e.g., Jobert et al., 2007; Musall and Kuik, 2015), in Japan (Maruyama et al., 2007) or the UK (e.g., Cowell et al., 2011). Multiple-choice experiments provide evidence that households prefer public compensation to private (e.g., Garcia et al., 2016; Langer et al., 2017) and compensation of some but not all community members can increase conflict (Baxter et al., 2013; Walker et al., 2014b), even be perceived as bribery (Gipe, 1995). Further, the receipt of direct individual compensation had a lesser effect on positive attitudes of wind turbine community members than perceiving the planning process as fair and just (Firestone et al., 2018; Hoen et al., 2019; Mills et al., 2019). Overall, these findings suggest that individual compensation can increase acceptance but is not a panacea to improve local acceptance (e.g., Jacquet, 2012; Hoen et al., 2019).

10.3.2 Participation and landscape perception

Wind turbines are visible, and their impact on the landscape often meets local opposition. Research on people–place relations and responses to place change suggests that this opposition can be understood as place-protective action (Devine-Wright, 2009). Recent research in the United States (Firestone et al., 2018; Hoen et al., 2019), the UK, and Norway (Batel et al., 2015) provide evidence that it is less the concept of place-attachment but of local and personal identity as well as the ideological social representation of a countryside. Batel et al. (2015: 149ff) found social representations to be used strategically: community members presented their area "as having more of the essence of the British or Norwegian countryside than other areas of the UK and Norway." Consequently, it can be argued that infrastructures like wind turbines or high-voltage transmission lines are even more distractive and displaced in this prototype beauty than in any other countryside or federal state. The concept of individual place attachment seems to narrow to capture the broader importance of symbolic meanings (e.g., Schöbel et al., 2013; Batel et al., 2015; Firestone et al., 2018). The symbolic meaning of landscape is mirrored in the significant impact of the wind project design on the local acceptance. For example, perceiving wind turbines as unattractive within the landscape post construction is negatively related to less-positive residents' attitudes but not seeing wind turbines per se (Hoen et al., 2019). The relevance of symbolic meaning is underlined by the finding that those residents "who do not like the look (26%) of their local wind project, 74% indicate it does not fit the landscape". Yet, similar percentages (73%, 78%, and 81%) indicate that the project is unattractive, industrial, and disruptive to the community feel. Interestingly, among those who like the look (63%), almost all (96%) indicate that it "symbolizes progress toward clean energy" compared to just 49% who indicate that "it fits well within the local landscape" (Firestone et al., 2018: 382). Thus, it is not visibility alone to influence attitudes but how wind turbines match the countryside identity. Citizens do have clear preferences for different layouts of offshore projects. If the design is connected to local features—such as historical meanings or the landscape—even larger projects are evaluated more positively (Hübner et al., 2018).

As the impact of visible infrastructure projects on the landscape is one of the major acceptance factors, visualizations of wind farms and high-voltage transmission lines are of crucial importance in participative planning processes. Visualizations are common at the level of permission processes but are a challenge at broader planning levels such as regional planning. At this early stage, potential siting areas are discussed—not specific projects. Consequently, visualizations could present scenarios only, and besides, the admission authorities are not empowered to have visualizations created. On the other hand, for community members, the visual impression of a potential wind farm is of crucial relevance. Opponents can use this lack of visual information to provide strategically, mere speculative, visualizations and stimulate negative reactions or at least uncertainty—even before a realistic project design is developed (Hübner et al., 2018). To avoid this problem even in the very early stages of a planning process, citizen participation should be integrated to discuss scenarios of areas suitable for wind energy. When community members have a say in the process and can influence a project's layout, the perceived fairness of the public process leading to approval is improved and more-positive attitudes toward a built project are likely (Firestone et al., 2018). However, although positive experiences with early and informal citizen participation exist (e.g., Devine-Wright, 2009; Rau et al., 2012; Rand and Hoen, 2017), it remains a challenge to transfer informal results into the formal planning process, competitive bid process, and the competitive strategy of wind developers to stake very early claims with single land owners.

It has been suggested before that seeing a wind turbine is related to negative attitudes, but this relationship was not significant when tested in a more complex acceptance model (see Hoen et al., 2019). However, the visibility of wind turbines by residences seems to increase annoyance (e.g., Pedersen and Persson Waye, 2007; Pedersen et al., 2009, 2010; Arezes et al., 2014; Firestone et al., 2015; Hübner et al., 2019a), as well as stress reactions (e.g., Pohl et al., 2018). Residents' financial interests in the wind project seem to decrease this negative impact (e.g., Pedersen et al., 2010; Arezes et al., 2014; Health Canada, 2014). The intercontinental comparison found that the number of visible turbines from the property was a significant predictor of noise annoyance stress in Europe but not in the United States. (Hübner et al., 2019a). These findings are relevant in regard to citizen participation as how positive and fair the planning process is experienced is strongly connected to noise annoyance by the operating turbines (Pohl et al., 2018; Hübner et al., 2019a).

10.3.3 Beyond participation—further factors

Citizen participation and community engagement are of crucial relevance. However, to foster the sustainable development of wind energy, further factors have to be addressed. To start with, the local acceptance of wind projects is influenced by the trust and credibility of local and national governments—governance is a recurring theme in Germany like in other countries (see Ellis and Ferraro, 2016).

Further, health concerns due to wind turbine sound can be solved by participation only partly but have to be addressed by reliable and valid empirical research and measures—comparable with how the impact on birds and bats is researched and mitigation approaches are developed. Indeed, due to residents' concerns, admission regulations have been adopted (e.g., restriction of shadow flickering duration on properties) and technical solutions have been developed to reduce the impact of shadow flickering, aircraft obstruction lights or to improve the noise quality of wind turbine sounds. While neglected until the last about ten years, meanwhile a first national health study was conducted in Denmark to gather reliable, scientific facts on the influence of wind turbine noise on residents, another large study on Health was conducted in Canada (Health Canada, 2014). Both studies did not find reliable relations between living in the neighborhood of wind turbines and health effects (Michaud et al., 2016a,b,c; Poulsen et al., 2018a,b). Still, while not connected to health problems wind turbine noise can evidently be annoying for residents and lead to less-positive attitudes (Haac et al., 2019; Hoen et al., 2019; Hübner et al., 2019a). A differentiated review on the impact of wind turbine noise is beyond the scope of the present chapter and is presented elsewhere (see Pohl et al., 2018; Hübner et al., 2019a). But it should be noted that several reliable studies on the impact of wind turbine sound on residents are available providing evidence that noise annoyance stress is negligibly connected to objective indicators, such as the distance from the nearest turbine—counterintuitively we even find more-positive attitudes for residents living closer than those living farther away (Baxter et al., 2013; Groth and Vogt, 2014; Hoen et al., 2019)—and sound pressure level modeled for each respondent but substantially with the perceptions of a lack of fairness of the wind project's planning and development process (Pohl et al., 2018; Hübner et al., 2019a,b). Beside the planning process, further subjective factors that seem to influence noise annoyance are visualized in Fig. 10.2, summarizing evidence from several studies (e.g., Pedersen and Persson Waye, 2004; Pasqualetti, 2011; Fast et al., 2016; Firestone et al., 2018).

Finally, wind turbine annoyance and related stress effects are not a widespread problem as the average annoyance levels of residents near wind farms in Europe and the United States were low, with the levels for noise similar across both samples, with European levels slightly higher for shadow-flicker, lighting and landscape change (Hübner et al., 2019a; Fig. 10.3). An international question seems to be, how this empirical knowledge can be disseminated effectively—and even more important, how trust in scientific facts can be restored.

Strongly annoyed residents present a small portion of the population with few differences between the United States and European annoyance stress levels (Hübner et al., 2019a).

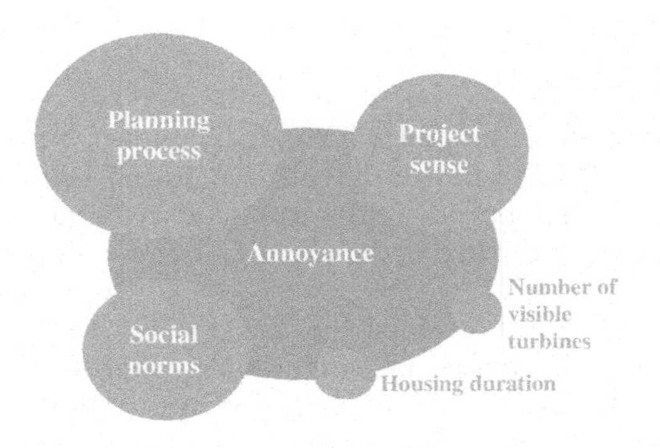

FIGURE 10.2 Social dimensions—the most important factors related to annoyance.

% (n) Total n	U.S. national survey	European sample
Sound	1.1% (16) 1441	4.3% (28) 657
Landscape change	1.5% (22) 1441	0.0% (0) 445
Lighting	1.2% (18) 1441	1.2% (10) 817
Shadow flicker	0.2% (3) 1441	0.2% (1) 445
Total	2.3% (33) 1441	3.7% (38) 1029

FIGURE 10.3 "Strongly" annoyance stress scale (inclusive symptoms).

10.4 Social acceptance of wind energy—the joint experience

Each nation and culture does have unique challenges but internationally comparable experiences with the social acceptance of wind energy projects and citizen participation prevail. Social acceptance has become a challenge even in the pioneer country of wind energy, Denmark. Throughout the industrialized, democratic societies, citizen participation is crucial but cannot guarantee positive attitudes of local wind projects, but without substantive citizen and community engagement, problems are more likely to increase. In fact, the relationships between residents' attitudes towards local projects, their perceptions about the particular fit of turbines within their landscape and community, health concerns and their perceptions of local benefits—and negative impacts—along with the perceived planning-process fairness and justice are complex. As social acceptance is a dynamic process with several layers of interactions, it is about social change, collective action, mobilization and popular rationalities (Batel et al., 2013). Due to this complexity and the international experiences, it seems unlikely that single fixes such as more community benefit funds or local citizen participation can solve the problem of the decreasing development of wind energy (Ellis and Ferraro, 2016). It seems that broader reforms are needed of how energy systems and the energy transition can engage with communities and citizens, inclusive implications for the national energy policies, ownership, institutions and regulation of wind energy and more innovative processes for engaging with citizens.

Although the larger question of social acceptance cannot be left to individual projects and their developers, developers are an important part of social acceptance. For example, the strong links between residents' experiences with wind farm planning processes, their levels of acceptance and even experienced stress impacts suggest to improve planning processes—such as by fostering opportunities to make voice heard that represent all groups of society, for example, the relative majority of supporters and the younger generation. In this respect, developers on their own can adopt proactive measures such as including neutral facilitators and, moreover, give participation the power to influence the projects features. However, it remains an open question how the last recommendation is compatible with competitive bid processes. Additionally, incorporating the social construction and local identity of the countryside in the design of wind projects may lead to higher acceptance, rather than focusing on turbine visibility alone (e.g., Schöbel et al., 2013; Batel et al., 2015).

It appears that wind turbines are considered good neighbors for the majority (e.g., Hoen et al., 2019). Communication strategies that make this positive experience public might maintain the public as well as local acceptance of wind projects. The same holds for the concerns of negative health impacts. Although links between health and exposure to wind turbines have not been found, this empirical knowledge does not seem to reach broader perception. This might be due to the fact that infra and low frequencies as well as the differences between noise quantity and quality can be mixed up easily (Pohl et al., 2018). However, some residents do experience annoyance and stress effects due to wind turbine noise, and therefore, health concerns might continue to be a major concern. Setback distances and limiting noise quantity impacts below the existing decibel restriction will not solve the problem as it is neither proximity nor noise quantity (when staying within the obligatory restriction) that are significantly connected to the stress annoyance of wind turbine sound but the noise quality and the perceived planning process. Further research on how to improve technical sound mitigating measures is recommended and annoyance, too, has to be understood in the broader perspective of citizen participation.

The important role of citizen participation, the role of governance and planning systems in social and local acceptance have been highlighted by a range of international empirical and practical experiences. The challenges of changing the energy transition to a low carbon economy seem to be a joined experienced in many respects—offering chances for joint solutions.

References

Aitken, M., Haggett, C., Rudolph, D., 2016. Practices and rationales of community engagement with wind farms: Awareness raising, consultation, empowerment. Plann. Theor. Pract. 17 (4), 557–576.

Albizu, L.G., Pagani, D., Brink, T., 2018. The Special Case of Denmark: April 2018, WWEA Policy Paper Series, PP-02-18-A.

Arezes, P.M., Bernardo, C.A., Ribeiro, E., Dias, H., 2014. Implications of wind power generation: exposure to wind turbine noise. Procedia Soc. Behav. Sci. 109, 390–395.

Batel, S., Devine-Wright, P., Tangeland, T., 2013. Social acceptance of low carbon energy and associated infrastructures: a critical discussion. Energy Policy 58, 1–5.

Batel, S., Devine-Wright, P., Wold, L., Egeland, H., Jacobsen, G., Aas, O., 2015. The role of (de-)essentialisation within siting conflicts: an interdisciplinary approach. J. Environ. Psychol. 44, 149–159.

Baxter, J., Morzaria, R., Hirsch, R., 2013. A case-control study of support/opposition to wind turbines: the roles of health risk perception, economic benefits, and community conflict. Energy Policy 61, 931–943.

Cowell, R., Bristow, G., Munday, M., 2011. Acceptance, acceptability and environmental justice: the role of community benefits in wind energy development. J. Environ. Plan. Manag. 54 (4), 539–557.

Cronin, T., Ram, B., Gannon, J., Clausen, N.-E., Thuesen, C., Maslesa, E., et al., 2015. Public Acceptance of Wind Farm Development: developer Practices and Review of Scientific Literature: Wind2050 WP3 Deliverable 1, DTU Wind Energy E, No. 0051.

Devine-Wright, P., 2009. Rethinking NIMBYism: the role of place attachment and place identity in explaining place-protective action. J. Community Appl. Soc. Psychol. 19, 426–441.

Ellis, G., Ferraro, G., 2016. The Social Acceptance of Wind Energy: Where We Stand and the Path Ahead. EUR 28182 EN.

Fast, S., Mabee, W., Baxter, J., Christidis, T., Driver, L., Hill, S., et al., 2016. Lessons learned from ontario wind energy disputes. Nat. Energy 1 (15028). Available from: https://doi.org/10.1038/nenergy.2015.28.

Fergen, J., Jacquet, J., 2016. Beauty in motion: expectations, attitudes, and values of wind energy development in the rural U.S. Energy Res. Soc. Sci. 11, 133–141. Available from: https://doi.org/10.1016/j.erss.2015.09.003.

Firestone, J., Kempton, W., 2007. Public opinion about large offshore wind power: underlying factors. Energy Policy 35, 1584–1598.

Firestone, J., Bates, A., Knapp, L., 2015. See me, feel me, touch me, heal me: wind turbines, culture, landscapes, and sound impressions. Land Use Policy 46, 241–249.

Firestone, J., Hoen, B., Rand, J., Elliot, D., Hubner, G., Pohl, J., 2018. Reconsidering barriers to wind power projects: community engagement, developer transparency and place. J. Environ. Policy Plann. 20 (3), 370–386. Available from: https://doi.org/10.1080/1523908X.2017.1418656.

Garcia, J.H., Cherry, T.L., Kallbekken, S., Torvanger, A., 2016. Willingness to accept local wind energy development: does the compensation mechanism matter? Energy Policy 99, 165–173.

Geißler, G., Köppel, J., Gunther, P., 2013. Wind energy and environmental assessments – a hard look at two forerunners' approaches: Germany and the United States. Renew. Energy 51, 71–78.

Gipe, P., 1995. Wind Energy Comes of Age. Wiley Press, New York.

Gross, C., 2007. Community perspectives of wind energy in Australia: the application of a justice and community fairness framework to increase social acceptance. Energy Policy 35, 2727–2736.

Groth, T.M., Vogt, C., 2014. Residents' perceptions of wind turbines: an analysis of two townships in Michigan. Energy Policy 65, 251–260. Available from: https://doi.org/10.1016/j.enpol.2013.10.055.

Haac, T.R., Kaliski, K., Landis, M., Hoen, B., Rand, J., Jeremy, F., et al., 2019. Wind turbine audibility and noise annoyance in a national U.S. survey: individual perception and influencing factors. J. Acoust. Soc. Am. 146 (2), 1124–1141. Available from: https://doi.org/10.1121/1.5121309.

Haggett, C., 2008. Over the sea and far away? A consideration of the planning, politics and public perception of offshore wind farms. J. Environ. Policy Plan. 10 (3), 289–306.

Health Canada, 2014. Wind turbine noise and health study: summary of results. <http://www.hc-sc.gc.ca/ewh-semt/noise-bruit/turbine-eoliennes/summary-resume-eng.php> (accessed 20.09.17.).

Hoen, B., Firestone, F., Rand, J., Elliot, D., Hübner, G., Pohl, J., et al., 2019. Attitudes of U.S. wind turbine neighbors: analysis of a nationwide survey. Energy Policy 134, 110981. Available from: https://doi.org/10.1016/j.enpol.2019.110981.

Hübner, G., Löffler, E., 2013. Wirkungen von Windkraftanlagen auf Anwohner in der Schweiz (Impact of wind turbines on residents in Switzerland). Institute of Psychology, Martin-Luther-University Halle-Wittenberg, Germany.

Hübner, G., Pohl, J., Schöbel, S., Kern, S., Gawlikowska, A., Marini, M., 2018. Akzeptanz Erneuerbarer Energien. (Acceptance of Renewable Energies). Institute of Psychology, Martin-Luther-University, Halle-Wittenberg, Germany.

Hübner, G., Pohl, J., Hoen, B., Firestone, J., Rand, J., Elliott, D., et al., 2019a. Monitoring annoyance and stress effects of wind turbines on nearby residents: a comparison of U.S. and European samples. Environ. Int. 132. Available from: https://doi.org/10.1016/j.envint.2019.105090.

Hübner, G., Pohl, J., Warode, J., Gotchev, B., Nanz, P., Ohlhorst, D., et al., 2019b. Naturverträgliche Energiewende – Akzeptanz und Erfahrungen vor Ort (Ecologically compatible energy transition – local acceptance and experiences). Institute of Psychology, Martin-Luther-University Halle-Wittenberg, Germany.

Jacquet, J.B., 2012. Landowner attitudes toward natural gas and wind farm development in Northern Pennsylvania. Energy Policy 50, 677–688.

Jami, A.A., Walsh, P.R., 2016. Wind power deployment: the role of public participation in decision-making process in Ontario, Canada. Sustainability 8 (8), 713.

Jobert, A., Laborgne, P., Mimler, S., 2007. Local acceptance of wind energy: factors of success identified in French and German case studies. Energy Policy 35 (5), 2751–2760. Available from: https://doi.org/10.1016/j.enpol.2006.12.005.

Kamlage, J.-H., Richter, I., Nanz, P., 2018. An den Grenzen der Bürgerbeteiligung (at the boarders of citizen participation). In: Radtke, J., Holstenkamp, L. (Eds.), Handbuch Energiewende und Partizipation (Handbook Energy Transition and Participation). Springer, Wiesbaden, pp. 627–642.

Langer, K., Decker, T., Menrad, K., 2017. Public participation in wind energy projects located in Germany: which form of participation is the key to acceptance? Renew. Energy 112, 63–73.

Maruyama, Y., Nishikido, M., Iida, T., 2007. The rise of community wind power in Japan: enhanced acceptance through social innovation. Energy Policy 35 (5), 2761–2769. Available from: https://doi.org/10.1016/j.enpol.2006.12.010.

Michaud, D.S., Keith, S.E., Feder, K., Voicescu, S.A., Marro, L., Than, J., et al., 2016a. Personal and situational variables associated with wind turbine noise annoyance. J. Acoust. Soc. Am. 139, 1455–1466.

Michaud, D.S., Feder, K., Keith, S.E., Voicescu, S.A., Marro, L., Than, J., et al., 2016b. Exposure to wind turbine noise: perceptual responses and reported health effects. J. Acoust. Soc. Am. 139, 1443–1454.

Michaud, D.S., Feder, K., Keith, S.E., Voicescu, S.A., Marro, L., Than, J., et al., 2016c. Self-reported and measured stress related responses associated with exposure to wind turbine noise. J. Acoust. Soc. Am. 139, 1467–1479.

Mills, S.B., Bessette, D., Smith, H., 2019. Exploring landowners' post-construction changes in perceptions of wind energy in Michigan. Land Use Policy 82. Available from: https://doi.org/10.1016/j.landusepol.2019.01.010.

Musall, F.D., Kuik, O., 2015. Local acceptance of renewable energy—a case study from southeast Germany. Energy Policy 39 (6), 3252–3260.

Pasqualetti, M.J., 2011. Opposing wind energy landscapes: a search for common cause. Ann. Asso. Am. Geogr. 101 (4), 907–917.

Pedersen, E., Persson Waye, K., 2004. Perception and annoyance due to wind turbine noise – a dose response relationship. J. Acoust. Soc. Am. 116, 3460–3470.

Pedersen, E., Persson Waye, K., 2007. Wind turbine noise, annoyance and self-reported health and well-being in different living environments. Occup. Environ. Med. 64, 480–486.

Pedersen, E., van den Berg, F., Bakker, R., Bouma, J., 2009. Response to noise from modern wind farms in the Netherlands. J. Acoust. Soc. Am. 126, 636–643.

Pedersen, E., van den Berg, F., Bakker, R., Bouma, J., 2010. Can road traffic mask sound from wind turbines? Response to wind turbine sound at different level of road traffic sound. Energy Policy 38, 2520–2527.

Pohl, J., Hubner, G., Mohs, A., 2012. Acceptance and stress effects of aircraft obstruction markings of wind turbines. Energy Policy 50, 592–600.

Pohl, J., Gabriel, J., Hübner, G., 2018. "Understanding stress effects of wind turbine noise – the integrated approach. Energy Policy 112, 119–128. Available from: https://doi.org/10.1016/j.enpol.2017.10.007.

Poulsen, A.H., Raaschou-Nielsen, O., Pena, A., Hahmann, A.N., Baastrup Nordsborg, R., Ketzel, M., et al., 2018a. Short-term nighttime wind turbine noise and cardiovascular events: a nationwide case-crossover study from Denmark. Environ. Int. 114, 160–166.

Poulsen, A.H., Raaschou-Nielsen, O., Pena, A., Hahmann, A.N., Baastrup Nordsborg, R., Ketzel, M., et al., 2018b. Long-term exposure to wind turbine noise at night and risk for diabetes: a nationwide cohort study. Environ. Res. 165, 40–45.

Rand, J., Hoen, B., 2017. Thirty years of North American wind energy acceptance research: what have we learned? Energy Res. Soc. Sci. 29, 135–148.

Rau, I., Schweizer-Ries, P., Hildebrand, J., 2012. Participation: The silver bullet for the acceptance of renewable energies? In: Kabisch, S., Kunath, A., Schweizer-Ries, P., Steinfuhrer, A. (Eds.), Vulnerability, Risks, and Complexity: Impacts of Global Change on Human Habitats, vol. 3. Gottingen, Hogrefe, pp. 177–191. Advances in People-Environment Studies.

Schöbel, S., Dittrich, A.R., Czechowski, D., 2013. Energy landscape visualization: scientific quality and social responsibility of a powerful tool. In: Stremke, S., van den Dobbelsteen, A. (Eds.), Sustainable Energy Landscapes. CRC/Taylor & Francis, Boca Raton, FL, pp. 133–160.

Walker, C., Baxter, J., Ouellette, D., 2014a. Beyond rhetoric to understanding determinants of wind turbine support and conflict in two ontario, Canada communities. Environ. Plann. 46 (3), 730–745.

Walker, C., Baxter, J., Ouellette, D., 2014b. Adding insult to injury: the development of psychosocial stress in Ontario wind turbine communities. Soc. Sci. Med. 133, 358–365.

Wiersma, B., Devine-Wright, P., 2014. Public engagement with offshore renewable energy: a critical review. WIRES Clim. Change 5, 493–507.

Wilson, G.A., Dyke, S.L., 2015. Pre- and post-installation community perceptions of wind farm projects: the case of Roskrow Barton (Cornwall, UK). Land Use Policy 52, 287–296.

CHAPTER

11

The contact group—public participation in the distribution network expansion in Baden-Württemberg

Regina Schroeter and Daniel Zirke

Netze BW, Stuttgart, Germany

OUTLINE

11.1 Introduction

Following the events in Fukushima in 2011, the German government has given a further impetus to the energy revolution in Germany. This does mean not only the renunciation of nuclear energy generation but also a desired socio-technical change culminating in three goals: (1) climate protection, (2) the increased expansion of renewable energies, and (3) the reduction in primary energy consumption (Fischer, 2011; Schroeter et al., 2016; Schröter,

The Role of Public Participation in Energy Transitions
DOI: https://doi.org/10.1016/B978-0-12-819515-4.00011-8

2017, 2018). The decision is generally supported by the majority of Germans, although the concrete implementation is controversial (Setton, 2017).

Individual measures throughout Germany are linked to the ambitious goals of energy system transformation. Between 2012 and 2017, the installed net capacity of wind turbines both on- and offshore almost doubled from 30.83 to 56.17 GW. The installed capacity in the solar sector also increased from approximately 33 GW in 2012 to over 42 GW in 2017 (Fraunhofer Institute for Solar Energy Systems, 2018). The extensive expansion of renewable generation facilities is also accompanied by a change in the generation landscape. Instead of less central, fossil-based large-scale power plants, numerous decentralized, regenerative plants are being built. This requires a change in the electricity grid infrastructure, which takes place at the transmission grid level (very high voltage) as well as at the distribution grid level (high, medium, and low voltage). In particular, the very high-voltage grid serves to transport electricity over long distances (e.g., from the offshore wind turbines in northern Germany to the high-consumption southern and western regions). In addition to the transmission grids, which are particularly present in the public, the distribution grids, which in the past primarily served the local and regional electricity supply, play an increasingly central role in the implementation of the energy system transformation. Today, 96% of all regenerative generation plants are already connected to the high-, medium-, and low-voltage grids. In this context can be seen that, despite the great support that the energy system transformation is receiving at a general level, concrete measures, such as the extension, conversion, and new construction of high-voltage lines, are received less positively by the population than would generally be expected on the basis of approval of the energy system transformation.

Against this background, new methods of informing and involving the public are increasingly being used. On the one hand, because an increasingly demanding public has clear ideas about the form in which they would like to participate in concrete planning projects—in this case the expansion of the network. On the other hand, also because in recent years, both general administrative regulations (e.g., § 25 para. 3 VwVfG [Administrative Procedures Act] and § 2 UVwG BW [Environmental Management Act Baden-Württemberg]) and specific energy regulations (e.g., §§ 12a para. 2 and 3, 12b para. 3 and 4, and 12c para. 3, 4, 6, and 7 EnWG [Energy Economy Act] and §§ 7 and 20 NABEG [Grid Expansion Acceleration Act]) on the use of early public participation have been adopted. These supplement the well-known and differentiated authorization law with the formal participation provisions contained therein (e.g., §§ 73 and 74 VwVfG, §§ 43a and 43b EnWG, and §§ 9, 10, 13, 22, and 24 NABEG). On the one hand, the aim is to inform the public about the projects and complex planning methodology. On the other hand, the methods are also intended to obtain planning relevant information from the people living on site and to respond to the interests of the individual participants as far as the framework conditions of the project allow.

Such a project, which envisages the construction of a new high-voltage line with a length of approximately 25 km, is currently being planned in Baden-Württemberg by the responsible grid operator (project sponsor).

To realize the new construction of the high-voltage line, two basic planning steps are required: the implementation of a regional planning procedure (cf § 15 ROG in conjunction with § 1 No. 14 RoV [regional planning procedure] and §§ 18, 19 LPlG BW [State Planning

Act Baden-Württemberg]) and a downstream planning approval procedure (cf §§ 43, 43a, and 43b EnWG in conjunction with §§ 72ff VwVfG). The first serves to review the project with the objectives and principles of spatial planning. The spatially significant effects of the planning are examined under supra-local aspects (i.e., the environmental and spatial compatibility of the project with regard to ecological, social, cultural, and economic requirements). The objective is to find a preferred spatially and environmentally compatible corridor with a width of up to 1000 m from various variants and alternatives. The second is then used as the approval procedure for detailed planning (i.e., the concrete determination of the line within the previously spatial corridor).

To meet the public requirements described above and to involve local people in the planning process as early as possible, the project sponsor carried out early public participation at the regional planning level. In addition to well-known participation methods (e.g., the organization of information events and citizen consultation hours as well as the presentation of the project in the press, on a project website, and in municipal and local councils), an innovative method, the contact group, was also used.

The following chapter describes the reasons that led to the selection of this method, its practical implementation, and the key results of the procedure. Finally, the results are critically discussed with regard to the objectives of the procedure as well as the experience gained.

11.2 Theoretical perspective and selection of the participation method

Corresponding to the objectives of the regional planning procedure described above, to select an appropriate method of citizen participation, the objectives of the participation procedure were first specified from three perspectives: First, participation should aim to provide participants with sufficient information. This was necessary because the project with its planning and approval process is multilayered and complex in terms of the technical, legal, environmental, and economic requirements. Second, in a logical step based on this, an attempt was made to acquire current and regional knowledge and incorporate it into the planning process. The concerns and wishes of the local public should also be taken into account. In this context, importance was attached to providing the participants with feedback from the project sponsor as quickly as possible on the extent to which recommendations of the contact group could be taken into account in the planning. Third, the overall objective was to win over the participants as multipliers in the region for the project, whereby it was not a question of agreeing with the positions of the project sponsor but rather of contributing to a constructive and relevant dialog (Ulmer and Sippel, 2016).

This view of public participation can thus be understood as functionalist from the outset. It emphasizes the integration of information into a decision-making process as a prerequisite for being able to intervene purposefully and successfully in increasingly complex socio-technical systems (Renn, 2008). The quality but also the effectiveness with which a decision can be made and implemented in a system depends strongly on synthesizing all information and value patterns associated with it in a participation process (Renn, 2008). From the perspective of functionalism, the provision of information at the beginning of a participation process is seen as a necessary condition for enabling participants to make meaningful contributions within the participation process. Because of the complex factual interrelationships, this understanding was also followed in the case of public participation within the framework of the regional planning procedure.

Admittedly, it can be argued against the interpretation that the regional planning procedure serves to weigh up all the relevant and level-appropriate interests brought into the project. Thus a neo-liberalist interpretation focusing on the weighing of different interests could also be considered. However, this interpretation confuses the formal regional planning procedure with the preceding early public participation, which itself is to be classified as a procedural step preceding the regional planning procedure.

If one starts from this functionalist self-understanding as the basis for the selection of a suitable participation method and follows the argumentation of Renn, participation methods involving a broad spectrum of different social groups in a decision are particularly important to get as comprehensive a picture of an object of decision as possible and interpret it as a functionalistically oriented participation procedure. In this context, the author proposes methods, such as the expert Delphi, public hearings, and citizen advisory committees (Renn, 2008).

Against the background of the above procedures, the contact group fits in well and, from a methodological point of view, comes close to the method used by the citizen advisory committee. According to the definition of the project sponsor, the contact group is a group of volunteers assembled, convened, and implemented under predefined conditions and selection criteria. With the help of a neutral moderation team, it actively accompanies and enriches the planning and approval process—within the framework of the defined mandate and the legal scope of action—for the period before and during the regional planning procedure. It functions as an informal instrument for early and nonformal public participation, whereby the participants of the contact group are free to participate in the formal participation procedure. The principle of voluntary action and constructive cooperation of all participants applies.

If one compares the method of the contact group with that of the citizen advisory committee, the two are very similar. Both methods involve a group of people meeting several times over a longer period of time. Participants are selected in such a way that they are representative with regard to relevant characteristics of the object of participation, whereas statistical representativeness cannot be achieved. Within the framework of both methods, the participants advise on complex infrastructure and planning topics, such as traffic planning or the expansion of the energy network. Furthermore, both methods make it possible to involve representatives of associations, federations, and interest groups (EPA, n.d.) in addition to private citizens.

Even though the 16-member contact group is somewhat smaller than typical citizen advisory committees, which usually have 12–25 members, and information can be provided in the contact group so that participants can comment on the relevant topics, the intersection between the two methods is large (EPA, n.d.). The contact group as a method can thus be assigned to the functionalist spectrum and appeared suitable before the stated objectives of communicating and exchanging information and multiplying the project sponsor's project.

11.3 Conception and implementation of the contact group

11.3.1 Definition of the group of participants

To achieve the goals described above, an interdisciplinary and socio-demographically balanced group of participants was required. In addition, the workability of the contact group had to be guaranteed. The contact group was therefore limited to a total of 16 participants. Eight participants came from regionally active associations, clubs, and initiatives (interest groups) that covered the topics of humans, nature conservation, environmental protection, forestry, agriculture, landscape and tourism, and business and culture, and thus a large proportion of those topics or protected interests were also the subject of the preparation of application documents by law (cf § 18 (2) LPlG BW and § 2 (1) UVPG [Environmental Impact Assessment Act]). The remaining eight participants were local nonorganized citizens. This was intended to ensure that, in addition to the partially professionally organized interest groups, which have extensive topic-related specialist knowledge, it was possible to contrast a broad social opinion to facilitate a multifaceted discussion. In addition, the contact group was supported by a moderation and documentation team. The project sponsor was also part of the group of participants. To introduce further interests and positions as well as expert knowledge to the contact group, it was also possible to invite guests. The participation of locally affected officers as contact group participants was not planned because they were already extensively involved in the planning process through other participation methods.

11.3.2 Participant acquisition

To achieve as balanced a group of participants as possible, two different approaches to attract the participants were chosen.

For the recruitment of representatives of the interest groups, it was first publicly announced via the press and the project website of the project sponsor that a contact group would be initiated. In addition, potential participants who were already known from a stakeholder analysis and through personal contact were contacted personally. This was accompanied by a request for an expression of interest if participation was sought. At the same time, two thematic areas that cover the content of the interest groups should also be identified. The selection of those interested in participating was then made by the moderation team for each topic lottery pot. If the same interested party was drawn for two different topics during the draw, a further draw took place because all interest groups were to cover only one topic at a time. The interest groups were then personally informed about the result of the participant selection.

To attract nonorganized citizens, information was also provided about the press and the project website of the project sponsor to initiate a contact group. On the one hand, this gave interested parties the opportunity to contact the moderation team or the project sponsor directly. On the other hand, random telephone calls were made via a renowned survey institute to acquire interested parties for involvement in the contact group. In a next step, the neutral moderation team conducted telephone interviews with all those interested in taking part in the project to determine the time available and to clarify open questions, in particular about the mutual expectation horizon. Any existing positions on the project were not relevant. However, it was important that at least a willingness to discuss and participate was signaled (Ulmer and Sippel, 2016). As a final step, a lottery pot procedure was planned from which eight participants could be acquired; these were personally informed.

All nonorganized participants were paid an expense allowance of €45 per contact group meeting (e.g., for travel expenses or necessary child care).

11.3.3 Mandate and functions

As already indicated in the objectives described above, the contact group had various immanent functions based on a previously defined mandate.

After that, the contact group first had an information function. It ensured that the contact group participants were informed by the project sponsor as well as guests and experts on various project-related topics and kept up to date on an ongoing basis (e.g., on parallel spatial and environmental studies conducted by the project sponsor). On the other hand, the contact group participants informed the project sponsor about the current public opinion.

Building on this, a second consultation function was established. This served to query the concrete local knowledge of the participants. It was not only important for the moderation team to identify the mere knowledge, opinions and positions but also the interests and needs behind them.

The third function, also based on this, provided for a defined codetermination of the contact group. The participants were then able to adopt concrete recommendations, which will be presented in detail later.

The fourth and final function of the contact group was to act as a multiplier. Accordingly, the contact group participants also acted as important contact persons, information providers, and recipients outside their involvement in the group (i.e., as semipermeable to the general public).

11.3.4 Implementation and contents

In total, the contact group met on eight dates between November 2014 and March 2015. All sessions were professionally prepared, accompanied, and documented by a moderation team. In most cases, the evening sessions were scheduled to last 3.5 h, although the desired end of the meeting was regularly overrun. This was due in particular to the complex situation, the high level of interest, and the resulting need for extensive discussion. Despite the time required and the complex topics, the attendance was very high over all meetings. On average, a maximum of two of the 16 participants were missing.

In terms of content, the meetings discussed in particular the reason for the plan (i.e., the necessity of the planned project), the corridors proposed by the project sponsor, and the potential technical design as underground cable or overhead line as well as the objects of protection affected by this—which had already been represented by the broad group of participants. For example, the impact of an existing military low-altitude helicopter corridor in space on the course of the corridor and the technological decision as to whether an overhead line can be used at all was part of intensive discussions. The affected settlements and the targeted distances were also discussed with regard to the corridor courses. Furthermore, the landscape and tourism, on which the project has potential impacts, were the subject of the meetings. The question of how much infrastructure expansion (especially the expansion of wind power plants, monocultures because of biogas plants and grid expansion) is reasonable for a region was also raised. To structure and constructively deal with these contents, some of which were emotionally documented, various other communication methods were used (e.g., simulation games and the awarding of points on the basis of the aforementioned objects of protection).

11.3.5 Transparency and results

The contact group generated two main results: a topic and recommendation protocol for each meeting and a concrete recommendation paper addressed to the project sponsor, the policy-makers, and the authority responsible for the procedure.

The topic and recommendation protocols, which were prepared at the end of each contact group meeting, reflect the main content of the meeting and discussion process. They also hold work orders to the project sponsor. All topic and recommendation protocols were published on the project website of the project sponsor so that the general public also had access to the topics and contents of the contact group. The aim was to achieve the greatest possible transparency despite the limited number of participants. This was strengthened by the opportunity to participate as a guest in the contact group as well as the opportunity to contribute topics, tips, and questions about the contact group participants, the moderation team, or the project sponsor. In addition, the regional press reported on the dates of the meetings so that additional insight was provided via this medium.

With regard to the mandate of the contact group, the final recommendations should also be emphasized as an important result. These were drawn up by the contact group at the meetings and adopted jointly at the last meeting. In accordance with the mandate of the contact group, these recommendations, including feedback to the contact group, were to be checked for feasibility by the project sponsor. Irrespective of their feasibility, it was also stipulated that the recommendations were to be enclosed with the application for a regional planning procedure to be prepared by the project sponsor so that the responsible regional planning authority was also given the opportunity to take account of the recommendations. In addition to the project sponsor's own interest in planning as transparently as possible, this type of documentation corresponds to the legal intention of early public participation and the documentation requirements contained therein (cf § 25 para. 3 sentence 4 VwVfG and § 2 para. 1 sentence 5 UVwG). In addition to the documentation in the application for the regional planning procedure, the recommendations were published in the local press, on the project website of the project sponsor, and at various information events.

11.4 Results from the contact group

The contact group's recommendations, which are the main outcome of the participation process, are addressed to the political stakeholders, the project sponsor, and the respective regional planning authority. The participants will notice in advance that the entire planning and planning process is assessed as complex and dynamic. The recommendations made may therefore adapt or overtake themselves in the further course of the underlying power line project.

In particular, there were calls for greater flexibility in the interpretation of standards vis-à-vis political decision-makers to be able to do even greater justice to regional specifics. It was also recommended that additional efforts by the state to clarify the necessity of grid expansion as a result of the expansion of renewable energies as well as research into alternatives to grid expansion be given increased support. Furthermore, it was recommended to adapt the support mechanisms in favor of a local energy turnaround (e.g., the strengthening of photovoltaics in urban areas).

It was specifically recommended that the project sponsor inform the public about the contents of the application before submitting the application documents for the regional planning procedure. In addition, it was demanded that the criterion of settlement distances be particularly highly weighted in the evaluation of the different corridor variants and alternatives. The recommendations also dealt with concrete planning information, among other things on nature and environment, landscape, monument protection, the handling of bundling areas, and the feared over-concentrations of infrastructure. Finally, in the recommendations, the contact group assessed the advantages and disadvantages of the corridors proposed by the project sponsor as well as the additional alternatives introduced in the contact group.

To what extent the results of the contact group were able to generate added value for planning, transparency and acceptance, which is inherent in early public participation, will be examined in Section 11.5.

11.5 Critical examination of the concept and results of the contact group

The discussions essentially can be understood as a contribution to an objective discussion of the project—both in the contact group itself as well as in the general public. Together with the participants, interesting and creative ideas (e.g., new corridor proposals) were introduced into the process. This information on planning and alternatives in planning led to an improvement in the quality of the application documents for the regional planning procedure because it was possible to take many arguments into account at an early stage of the application process. The arguments also had some influence on the evaluation and selection of the corridors. The early discussion made it possible to intensively prepare numerous topics, which would otherwise be the subject of the regional planning procedure. It remains to be seen to what extent this will result in a discharge of the formal procedure because the regional planning procedure has not yet been completed. There is still a risk that certain issues will be discussed twice. All in all, the mixed field of participants in the contact group enabled a balance to be struck between private citizens and organized interest representatives. This was reflected in the content of the argumentation to the effect that, in addition to the interests of nature conservation and environmental protection, which were otherwise strongly emphasized by law in planning procedures, private interests might be juxtaposed.

Despite this positive balance of the discussion, the complexity and variety of the topics and questions within the framework of the project proved to be a particular challenge for the participants. Regardless of the mandate of the contact group, in the first three meetings, for example, the question of whether the project should be implemented at all (Necessity of planning) was mainly debated. The discussion on how the project should be implemented also sometimes led to deviations in the mandate of the contact groups. Even at this early stage of planning, certain topics were addressed in great detail, although the content of these topics can only be assigned to the subsequent planning approval procedure. Among other things, the technical design of the high-voltage line (whether underground cable or overhead line) or the exact course of a potential later route were discussed. Such inputs were taken for the subsequent level of planning in a thematic memory. Furthermore, some of the alternative corridor proposals, which go beyond the study area, reflect the NIMBY (not in my backyard) effect, although the reasons given are understandable.

In general, the contact group was well received by the public. Nevertheless, transparency in the provision of information to the public fell short of expectations, although the results of the contact group meetings were conveyed through minutes, the press, and the multiplier function of the participants. On the one hand, although the contact group brought together very different participants, it was representative in neither the qualitative sense nor the quantitative sense. On the other hand, during the meetings of the contact group, the project was partly discussed very emotionally in public. It therefore seems questionable how pronounced the individual members of the contact group were able to perform their multiplier functions.

Based on the experiences of the contact group, it should be considered whether the participation procedure should be carried out with several groups and on a broader participant base. On the one hand, this makes it easier to ensure the qualitative representativeness of the procedure; on the other hand, it may also make it possible to discuss different topics with different participants. This would reduce the effort involved in the procedure for individuals. In a second step, the results were then brought together in a plenary session. This somewhat larger event could generate greater public attention, which, in turn, could contribute to the transparency of the procedure.

Furthermore, experience has shown that a planning graded procedure, which is realized by the regional planning procedure or the subsequent planning approval procedure, can only be methodically represented to a limited extent in a participation procedure. Even information provided at the outset can only counteract this problem to a limited extent. Nevertheless, it seems necessary to comprehensibly present this general information on the planning procedure and its various levels at the beginning of the participation procedure. This makes it possible for as much information as possible to be adequately taken into account in the respective planning step and prevents frustration on the part of all parties involved.

Finally, experience shows that the protection of participants should not be underestimated, especially against the background of emotional debates. Here, it is appropriate not to hold the individual meetings in public and (as already happened in the case described) to publish the results of the contact group anonymously. The individual participants can remain free to express their opinions in public.

References

EPA (Environmental Protection Agency), n.d. Public Participation Guide: Citizen Advisory Boards. <https://www.epa.gov/international-cooperation/public-participation-guide-citizen-advisory-boards> (accessed 08.01.19.).

Fischer, S., 2011. Außenseiter oder Spitzenreiter? Das "Modell Deutschland" und die europäische Energiepolitik. Aus Politik und Zeitgeschichte: Ende des Atomzeitalters. 61, No. 46 – 47/2011, November 2011, pp. 15–22.

Fraunhofer Institute for Solar Energy Systems, 2018. Installierte Netto-Leistung zur Stromerzeugung in Deutschland. <https://www.energy-charts.de/power_inst_de.htm> (accessed 08.01.19.).

Renn, O., 2008. Risk Governance. Coping with Uncertainty in a Complex World. Earthscan, London/Sterling.

Schroeter, R., Scheel, O., Renn, O., Schweizer, P., 2016. Testing the value of public participation in germany: theory, operationalization and case study on the evaluation of participation. Energy Res. Soc. Sci. 13, 116–125.

Schröter, R., 2017. Beteiligungsverfahren zwischen Inklusion und Konvergenz: ein analytisch- deskriptives Modell. <https://elib.uni-stuttgart.de/handle/11682/9675> (accessed 03.09.18.).

Schröter, R., 2018. Erdkabel oder Freileitung? Stromnetzausbau zwischen kommunaler Politik und Öffentlichkeitsbeteiligung. <https://www.netzwerk-buergerbeteiligung.de/fileadmin/Inhalte/PDF-Dokumente/newsletter_beitraege/3_2018/nbb_beitrag_schroeter_181025.pdf> (accessed 08.01.19.).

Setton, D., 2017. Soziales Nachhaltigkeitsbarometer der Energiewende 2017. <http://publications.iass-potsdam.de/pubman/item/escidoc:2693915:8/component/escidoc:2734897/2693915.pdf> (accessed 03.09.18.).

Ulmer, F., Sippel, T., 2016. Frühzeitige Öffentlichkeitbeteiligung am Beispiel eines geplanten Neubaus einer 110-kV- Hochspannungsleitung. <https://dev.netzwerk-buergerbeteiligung.de/fileadmin/Inhalte/PDF-Dokumente/newsletter_beitraege/1_2016/nbb_beitrag_ulmer_sippel_160321.pdf> (accessed 08.01.19.).

CHAPTER

12

Social sustainability: making energy transitions fair to the people

Daniela Setton

Institute for Advanced Sustainability Studies (IASS), Berliner Strasse, Potsdam, Germany

OUTLINE

The Role of Public Participation in Energy Transitions
DOI: https://doi.org/10.1016/B978-0-12-819515-4.00012-X

12.1 The social dimensions of energy system transformation

12.1.1 Energy systems as sociotechnical systems

"To move away from the conventional, purely into renewable energies and more energy efficiency" (BMWi, 2017: 4) is how the German Federal Ministry of Economics and Energy describes the "core" of the energy transition. Accordingly, the objectives set by the federal government for the energy system transformation are primarily quantitative targets such as reduction rates for greenhouse gas emissions, the phasing out of conventional energies such as coal and nuclear power, growth rates for the share of renewable energies with respect to gross electricity, and heat consumption combined with absolute reductions in final energy consumption and growth rates for greater efficiency (BMWi, 2016: 8). In this respect, the federal government sees the challenges to achieve these goals above all as a technological challenge (Grasselt, 2016). This is revealed by a traditional understanding of changes in energy systems as a change in supply sources and the associated technologies, which is also widespread in the public debate on energy system transformations (Miller et al., 2013). The focus is on technological innovations and their diffusion, such as the expansion of renewable energy plants, in particular solar and wind energy, new efficiency and storage technologies, or the expansion of transregional power grids, in combination with the necessary promotion instruments and regulatory adjustments. In energy research, too, the focus is strongly on technological problems and solutions (D'Agostino et al., 2011; Sovacool, 2014a; Geels et al., 2017a,b; Sovacool 2014b).

The widespread understanding of energy system transformation as a technological and economic challenge is insufficient for gaining a full understanding of the complex challenge at hand. Energy systems are sociotechnical systems, which do not only include machines, plants, and technical devices but also the development, production, and use of technologies by individuals in society (Miller et al., 2013: 136). The application of energy technologies is closely linked to appropriate social practices, norms, ways of thinking, institutions, as well as models of work and knowledge. The spatially and temporally specific nexus of technical and nontechnical elements is also central to the functionality of the energy system (Grunwald et al., 2018: 831). In this respect, transformations of energy systems need to be understood more comprehensively as changes in sociotechnical systems, in which the expansion of technological innovations corresponds with social processes of change, for example with regard to consumer and investment behavior, power relations, political disputes and voting behavior, certain narratives, habits, lifestyles or worldviews, employment relationships (Miller et al., 2013, 2015; Miller/Richter, 2014; Schippl et al., 2017; Geels et al., 2017a,b).

The technological and social aspects of the change of energy systems must therefore be considered in an integrated way (Grunwald et al., 2018). Applying the analytical concept of the social dimensions of energy transformations, the historically specific processes of societal change accompanying technological change can be identified and examined. The concept refers to the social processes that drive and shape energy systems. Furthermore, it incorporates the social changes that trigger feedback loops to energy technologies, and the social consequences that result from the organization and functioning of new energy systems (Miller et al., 2013: 136).

In this respect, new technologies are necessary but not sufficient conditions for the success of energy transformations (Grunwald et al., 2018). Miller et al. (2013: 139) emphasize that the central decisions that have to be taken in the context of energy transformations do not relate to the various energy sources, but to the various forms of social, economic, and political arrangements that all together determine the changes in the energy system. The real challenge, then, is to create a new sociotechnical context in which the goals of the energy transitions can be met.

12.1.2 The German energy transition as a social-ecological transformation

This challenge is particularly demanding in the socio-ecological transformation of energy systems toward sustainability. Since it demands a comprehensive transformation of the entire energy system from fossil and nuclear to renewable energies, oriented toward the prevention of climate change, it differs considerably from the energy-related structural changes of recent decades (Czada and Radke, 2018; Czada 2014; Geels et al., 2017a,b). First, with regard to the far-reaching technological change, it requires an almost complete replacement of the dominant fossil energy sources and the change from a central conventional to a more decentralized renewable system. Second, it can only be successful if it is accompanied by "cultural, social, (. . .) economic, infrastructural as well as production- and consumption-related co-evolutionary changes in various sectors and systems of society" (Grießhammer and Brohmann, 2015: 5).

In view of the time pressure for immediate action to meet the climate policy challenge, the energy transition must also be promoted and accelerated politically, for example, by changing the regulatory framework, providing more economic incentives and enhancing the development and expansion of climate-friendly technologies (Geels et al., 2017a,b). In its latest special report (IPCC, 2018), the International Panel on Climate Change made it clear that the existing climate protection commitments under the Paris Climate Change Convention are not yet sufficient to produce the radical reduction in greenhouse gas emissions required to meet the 1.5°C target by 2030. The Paris accord calls for a rapid social transformation toward a climate-neutral economy and a new fossil-free way of life. Not only small, gradual changes are insufficient for this, but fundamental political changes are required, which imply far-reaching political measures (Heyen and Brohmann, 2017). As a social-ecological transformation toward sustainability, the energy transformation in

Germany is "subject to the more radical claim of a comprehensive collective design" (Brand, 2017: 27).

12.1.3 Monitoring of social dimensions necessary

To meet the challenges of the energy transformation, it must not only be mastered as a technological change but also as a "socio-cultural transformation challenge" (Wuppertal Institute, 2016: 10). It is not enough to "develop new technologies and launch new smart system solutions." Rather, the success of the energy transition will be measured by how it will be possible to combine what is technically possible with what is socially desirable (Renn and Dreyer, 2013: 41). How this can be achieved in concrete social practice is one of the central questions for the success of this transformation process.

What is needed is a solid knowledge base on what the "social" solutions for the transformation of energy systems could look like, so that social change can proceed as quickly as possible in the context of energy system transformation. Due to the relevant focus here on the social dimensions of energy transformation toward sustainability, we define socially desirable as what is actually desired in society. The necessary social changes for the success of the energy transformation will not be achieved if they are not largely supported politically by the population and actively implemented by those who have to make the behavioral changes required for the transition to become effective. What is socially desirable can therefore only be determined empirically by investigating the preferences, values, and behavioral intentions of individuals and groups in society.

With the fundamental reorganization of the energy system toward climate change, energy efficiency, energy saving, and renewable energies, many areas of people's lives are also deeply affected: from living conditions to work, mobility, consumption, and leisure activities. Not only is the acceptance (Wüstenhagen et al., 2007) of technologies such as wind and solar power plants crucial for progress. A high degree of political approval is also required for far-reaching political reforms that affect broad sections of the population in their everyday lives, such as a change in pricing and supply patterns in the energy sector. It is also necessary that the population itself actively promotes the transformation of the energy system, that is, buys more efficient technologies, replaces old, inefficient household appliances with more efficient ones, insulates houses, changes mobility behavior and refrains from using combustion engine for mobility, purchases its own solar systems and storage tanks, etc.

It is therefore essential that the associated changes are perceived positively or at least are tolerated. New practices and consumption patterns need to be learned, which presupposes that there are sufficient incentives, that the necessary knowledge is imparted and that there are sufficient practicable opportunities for behavioral changes in everyday life.

If large sections of the population have neither the need nor the willingness to invest in their own renewable energy systems or do not see any opportunities for participation, the necessary changes will simply not take place, at least under the rules of democratic regimes. This fact needs to be taken into account when making political decisions and choosing instruments and measures to promote energy system transformation and climate protection. This is also true for ideas of justice or for dealing with conflicts within the framework of the energy transformation. It is helpful to know to what extent potential conflicts are perceived and how individuals and groups prefer to resolve conflicts and to deal with trade-offs. A solid knowledge base on the values, preferences, and concerns of the affected populations is indispensable for recording the current status and the latest developments and trends and for drawing the appropriate conclusion when designing policies and communication programs for facilitating the energy transition.

What the necessary social arrangements should look like, for example, in the transport sector or in the use of smart technologies in private households, is a decisive question that will have a major impact on how the transformation process is going to be shaped. Continuous monitoring is crucial here. The German Advisory Council on Global Change (WBGU) has described the transformation to a climate-friendly society as an open social "search process" (WBGU - German Advisory Council on Global Change, 2011: 23), which needs orientation by the provision of knowledge but also needs the inclusion of social values and preferences for being politically and socially feasible. An empirically sound knowledge base on the attitudes, experiences, concerns, and preferences of the population toward the energy system transformation is an important prerequisite for a reflexive governance of the energy system transformation (Renn, 2015a: 68). Empirical studies that show to what extent public attitudes and concerns indicate developments leading toward sustainable or nonsustainable futures need to be incorporated into the political discussion and decision-making processes.

While there is a largely secure scientific knowledge base when it comes to determining the "technically possible," this can only be said to a limited extent in the case of the "socially desirable." Even though there are increasingly social and political science publications on the acceptance and social compatibility of the energy transition that address the issues of acceptance and social compatibility with regard to low-carbon transitions (Holstenkamp and Radtke, 2018; Gaede and Rowlands, 2018; Fraune et al., 2019), there are still clear gaps in knowledge. Only very few studies deal with the survey of the attitudes and preferences of the population with regard to the energy transformation across topics and on a long-term and systematic basis.

With the "Energy of the Future" process, the German Federal Government is already carrying out regular monitoring of the achievement of the quantitative objectives and the status of implementation of the energy transition measures, which to date has been based in particular on the available energy statistics (BMWi, 2016, 2019). However, the social dimensions of the energy system transformation have so far only been taken into account to a very limited extent with regard to employment development or the affordability of energy for private households. In its latest progress report on the implementation of the energy transition (BMWi, 2019), the Federal Ministry of Economics and Energy also emphasizes the importance of the acceptance of the energy transition by the population for the success of the transformation process. In its last two written statements, the independent expert commission on monitoring appointed by the federal government stressed the increasing importance of acceptance monitoring (Löschel et al., 2019). However, it is currently not evident that the ministry is adequately taking into account the social dimensions of the energy transition to continuously measure where we stand in Germany in this respect.

It is urgently necessary to close this gap in the monitoring of the German energy system transformation so that an analysis of the current status and the consequences can be carried out with regard to the social change associated with the transformation process , and so that needs for action can be identified at an early stage and undesirable developments corrected.

But how would such monitoring have to be set up to provide a sound knowledge base? What needs to be looked at more closely and measured to observe the social dimensions of the energy transitions in the long term? How can it be selected what is "socially" relevant in the transformation of the energy system transformation? This is where the concept of social sustainability comes into play. It provides the conceptual basis for a problem-oriented and normatively founded research perspective on the social dimensions of the energy system transformation.

This creates a focus for the monitoring of the energy transformation, which at the same time makes it possible to take a systematic look at important social prerequisites, implications, and consequences of the transformation process that have so far been widely neglected in the discussion on energy transitions. The operationalization of this concept by selecting suitable indicators for empirical studies will make developments in the social dimension of the energy transformation "measurable" and thus also assessable for policy making.

To make the concept of social sustainability useful in terms of recording and evaluating the social challenges of the energy transition, a new, application-oriented approach is needed, which is described below. First, a model of a socially sustainable energy system transformation will be sketched, followed by a description of how this model is transformed into an empirical design with indicators.

12.2 The concept of social sustainability of the energy transformation

12.2.1 Social sustainability: a contested concept

There is a broad consensus in the literature that sustainability is a normative concept for the future development of society, which refers to intra- and intergenerational justice. The focus is on the "distribution of opportunities and resources in comparison of peoples and individuals within the population living today (intragenerational justice) as well as long-term responsibility towards future generations" (Renn et al., 2007: 9). At the heart of the sustainability concept are thus globally and permanently viable live styles and routines for doing business (Ekardt, 2016). However, opinions differ widely as to what this normative goal means in concrete terms and how to implement intra- and intergenerational justice in the various social contexts. Furthermore, there is no agreement in the literature about which indicators are to be used to measure and assess the current state of sustainability in a given society.

Furthermore, the social dimension of sustainability has not yet been at the forefront of the sustainability debate (Ritt, 2002). There is now a whole range of studies on the topic, including very different theoretical approaches, thematic references, and contexts of use, as well as correspondingly diverse understandings of terms (see for example, Opielka, 2017; Jahrbuch Ökologische Ökonomik, 2007; Littig and Grießler, 2005; Empacher and

Wehling, 2002; Ritt, 2002). However, the concept of sustainability was already inspired by a socially influenced model as manifested in the definition of the Brundtland Report "Our common future" of 1987 and in the resolutions of the United Nations Conference on Environment and Development (UNCED) in Rio de Janeiro in 1992 (Empacher and Wehling 2002; WECD, 1987).

While there is now broad agreement in the debate that sustainability can only be achieved by integrating the various dimensions such as ecology, economy, and social affairs, the question of its definition and conceptual delimitation remains controversial. Most approaches have in common that they base their understanding of the social dimension on the so-called three- or multi-pillar model, in which the social area, including justice, participation, or peaceful conflict resolution, is regarded separately from the dimensions of economic practice and ecological impacts. This separation is also kept when it comes to specifying normative and analytical aspects of each pillar (Kopfmüller et al., 2001: 47ff; Michelsen and Adomßent, 2014: 23ff). Some approaches have developed integrative concepts of sustainable development where the various sustainability dimensions are dealt with in its interlinkages (Rösch et al., 2017).

Furthermore, the criteria and goals of social sustainability are often defined as generally desirable social goals or conditions that relate both to the macrolevel, that is, "to the preservation and reproduction of the core system of a society" (Opielka, 2017: 19), and to the microlevel, that is, the living conditions of individuals (Empacher and Wehling, 2002). For example, one can find numerous aspects that all relate to social sustainability in the discourse. Among them are social justice, distributive justice, cultural integration, social cohesion, social peace, social resources, acceptance, social compatibility, participation, employment, education, equal opportunities, quality of life, a "good life," securing health or satisfying basic needs. In a narrower understanding of the term, the "social" aspect of sustainability is largely limited to the questions of redistribution of resources, wealth, and opportunities. In a broader concept, social sustainability is also understood as an umbrella concept of the sustainability in general and thus as a "complex concept that ranges from methodology to strategies to value patterns" (Opielka, 2017: 28).

12.2.2 The application-oriented approach

The question of whether developments in the context of the energy transformation are going in the desired "socially sustainable" direction cannot be answered by "inventing" or selecting an ad hoc normative concept and deducting indicators on the basis of this concept. There is, first, a need for contextualization of the definition within the local and temporal conditions and, second, for a systematic approach incorporating the well-established theories of social change in the social sciences.

This chapter is informed by the existing theories of social change and transformation but its focus is on the first condition: placing social sustainability in the context of a local and temporal situation: the energy transition in Germany. We will not delineate a general concept or theory of social sustainability nor produce a comprehensive sustainability assessment of the energy system transformation which has been done by other scholars (such as Rösch et al., 2017). Our goal is to develop an application-oriented concept specifically related to the energy system transformation in Germany that can serve as a basis for monitoring the social dimensions that are crucial for making the transition move into the direction of sustainable practices. This task requires a special approach. The "social" aspect of sustainability is not defined on the basis of fundamental social-theoretical considerations or with regard to generally "desirable social conditions," as is usual in many conceptualizations of social sustainability but as a pragmatic still systematic search for criteria that make the transition to a sustainable energy future more likely to happen.

This is illustrated briefly by the following example. If abstract social sustainability norms or rules, such as social justice, are applied to energy system transformation and forms the basis for the indicator development, one could study how different and abstract concepts of social justice might face more or less support in the society. From a transformative or action-oriented perspective (facilitating transitions), this approach would not be very useful for providing more detailed insights into concrete ways of facilitating social solutions for the envisioned change. Therefore we have to consider the specific context of the transformation process at stake already at the concept development stage. We like to investigate what our respondents in the respected populations perceive as just or unjust in reference to a specific or concrete transition process. Therefore we gain insights on what is more likely to be desired given their attitudes and the justifications that they provide. This can only be determined empirically by "measuring" what the population thinks and feels, what individuals and groups desire and worry about with regard to concrete political measures, reform proposals or developments of the energy system transformation.

Central to this perspective is also the reference to permanence and global viability in the sustainability issue (Ekardt, 2016: 50). Already at the conceptualization stage, we refer to "social sustainability" in a temporally and spatially concrete societal context, in which the central issue is coping with the climate crisis, to the energy transition in Germany and the context of sustainable transformation.

12.2.3 Integration of normative and functional aspects

Our approach has been inspired by the normative-functional sustainability concept proposed by Renn et al. (2007). This concept includes systemic integrity, justice, and quality of life as the three basic norms of sustainability. The criteria and indicators are derived from the classic functional requirements developed by the structural-functional school in social theory (production, order, meaning, and social cohesion). The 2007 concept provides a detailed justification for using this basic theoretical foundation.

Based on this foundation, our concept encompasses the social sustainability of lifestyles and ways of doing business that are also instrumental for achieving the ecological target dimensions envisaged by the energy system transformation. The necessary social transformation process can only succeed within the framework of democratic processes and the social propensity of individuals and groups to accept changes in everyday routines. Only "if the social order is perceived as just, does it have a chance of long-term survival" (Renn et al., 2007: 59).

Taking into account the functional requirements for a successful energy system transformation, our concept of social sustainability formulates a normative perspective on how the social change processes that prove to be necessary to achieve the agreed-upon ecological objectives should be designed. A socially sustainable energy system transformation is not only accepted by a large majority of the population but also actively supported through changes of daily routines and involvement of many households in providing energy services (prosumer) and participating in community energy decisions. Social sustainability thus defines a positive model for the energy system transformation as a "joint effort for society as a whole" (Ethics Commission for Safe Energy Supply of the Federal Government, 2011), in which all parts of society feel addressed and can actively contribute. Accordingly, for an energy transformation to be called socially sustainable, it is required that its design and development is in line with the concerns, needs, and preferences of a broad majority of the population. This is both a normative and a functional requirement for obtaining a socially sustainable energy transformation.

In this respect, the social dimension is not lower ranking but of equal importance as the ecological or economic dimension of sustainability. This is more than just securing acceptance for necessary interventions to improve the ecological situation. The more all groups of society feel positively addressed and actively involved in the energy system transformation as a "joint effort" the more likely it will be to win the political majorities necessary for the concrete transformation and to initiate changes in the behavior of broad sections of the population.

12.2.4 Crucial criteria: participation and acceptance

A socially sustainable energy transformation is based on a high extent of societal participation of broad sections of the population, not only in political and economic terms but also with regard to the life chances of individuals (Gabriel and Völkl, 2008; Holtmann, 2015). Participation ranges from an appropriate level of energy supply for all people, a distribution of costs and burdens that are perceived as fair, inducing behavioral changes in everyday life or co-financing investments in renewable energy plants. The guiding principle of a socially sustainable energy transformation implies that negative effects on certain sections of the population are either avoided, kept to a minimum or mitigated by additional measures. Social compatibility with people's preferences and requests is a central component of social sustainability. Equally important are aspects of acceptance (Wüstenhagen et al., 2007; Renn, 2015b; Upham et al., 2015) and political participation (economic/political) (Holstenkamp and Radke, 2018). All in all, there are three dimensions of participation that are central to the energy system transformation: political, economic, and individual life opportunities (Table 12.1). Justice issues play a central role in all three dimensions.

TABLE 12.1 The three social sustainability dimensions of energy system transformation.

Political	Social political acceptance of policies, local acceptance of wind- and solar and other infrastructure expansion; participation in decision-making processes.
Economic	Economic and financial involvement—consumer—and investment behavior (market acceptance), distribution of costs and benefits.
Individual life opportunities	Social acceptability, personal experiences (supply with energy, renewable energy expansion); expectations on key aspects of individual live (employment, financial, and economic situation, etc.).

12.3 The social sustainability barometer for measuring social resonance to energy transformations

12.3.1 Focus on the concerns of the people

When it comes to assessing the social sustainability of energy transitions, the population itself must be placed at the center of attention. In the concept presented here, the benchmark for assessment lies with the people. It is about how social development is perceived, experienced, and evaluated by the people and what their preferences are. The more positive people's attitudes, assessments, and experiences of the energy system transformation are the better in terms of social sustainability. To measure social sustainability empirically, the population must be surveyed.

Nevertheless, this normative goal is linked with functional requirements such as security of supply, assurance of democratic processes, and social compatibility with basic values and worldviews. Whether these functional requirements are met, is a question of empirical research. It is essential for meeting these requirements to investigate how people perceive experience, value, and make judgment about energy-related changes in their lives. This requires empirical research methods such as surveys, focus groups, or experimental designs.

So far, only limited empirically sound knowledge is available on how the German population or certain groups perceive and evaluate the energy system transformation. There is also little knowledge about the extent to which people like to participate in both the individual transformation toward sustainable energy practices and the collective decision-making process about community infrastructure or political regulation. It is also knowledge missing on what ideas of justice and preference about distributive fairness people share when it comes to energy systems.

The empirical significance of many surveys on the subject of energy system transformation, mostly commissioned by associations, the media (television stations, magazines or newspapers) or companies, varies greatly. There are a large number of individual findings available, which are hard to compare. In addition, many surveys commissioned by associations or companies are carried out on an ad hoc basis and serve political interests. There are only few studies conducted by scientist that systematically approach the question of how the population perceives and evaluates the energy transition (Schumann, 2017; Ziegler, 2017; Groh and Ziegler, 2017; Sonnberger and Ruddat, 2016; BMUB/UBA, 2017; Saidi, 2018). However, even these scientific studies do not yet provide sufficient data for the monitoring of the three social dimensions that were discussed above. First, the thematic range of the available data is limited in scope and not covering all three dimensions. Second, only very few of the surveys are conducted in regular intervals and thus are unable to record changes or trends over time.

12.3.2 Measuring social sustainability

Against this background, the Institute for Advanced Sustainability Studies (IASS) has developed a new empirical instrument, the so-called social sustainability barometer which is published annually and was presented for the first time in 2017 (Setton et al.,

2017; Setton, 2019a,b). Based on our concept of social sustainability, it provides an empirical data basis for the long-term observation of the three social dimensions of energy transformations. The social sustainability barometer measures the three dimensions and their interactions with respect to the German energy transformation process and focuses on relevant action fields and aspects of the transformation process. For this purpose, a new, partly flexible indicator system has been developed with which attitudes, perceived justice and fairness, preferences, experiences, and expectations of individuals and groups are collected annually. The results are published in different formats one for policy makers and the other for academic audiences.

The social sustainability barometer uses a mixed-method approach: quantitative and qualitative research approaches are combined to address the complex research questions more comprehensively. At the core of the barometer is an annual Internet-based, population-representative household survey of over 6500 households in Germany, which has been conducted three times (in summer 2017 and 2018, and in autumn 2019) in cooperation with the RWI-Leibniz Institute for Economic Research as part of the forsa.omninet household panel. The selection was made using a multi-stage random selection procedure. The evaluation of the household survey data was based on statistical and explorative analyses. In addition to the household survey, the research team conducted qualitative in-depths interviews with relevant experts, focus group discussions with selected populations, and three citizen forums (Matuschke and Renn, 2019).

The Barometer has been developed in the context of the German-based dynamis partnership and in cooperation with the Kopernikus Project ENavi financed by the Federal Ministry of Research. Dynamis, a self-proclaimed German "think-do-rethink" tank, was founded in December 2016 by the innogy Foundation for Energy and Society, the 100 prozent erneuerbar Fondation, and the IASS.

The results of the barometer are both descriptive and normative. The barometer serves as a tool for describing the status quo and tracking positive and negative developments. The data it contains also highlight the existing or emerging challenges and problems. At the same time, however, they also serve as orientation for policy makers when trying to design policies and communication programs. The Barometer's findings point to areas where there is a need for political action and thus serve as an early-warning system to help policy makers to set the right priorities. The indicators do not aim to monitor the success in achieving specific quantitative policy targets like a share of renewable energies in gross electricity consumption as is the case of the federal government's monitoring process "Energy for the Future." Rather, it is a matter of "observing developments and assessing whether they are going in the desired direction, to be able to take countermeasures if necessary in the opposite direction" (Empacher and Wehling, 2002: 57). For this reason, no target values for the indicators are formulated here. Which implications the results should have for the political process is addressed in a follow-up activity of the IASS to engage with stakeholder and policy makers to "make sense" of what the studies demonstrate and to delineate in a joint discourse the necessary lessons for action. The barometer is based on transdisciplinary and transformative research (Nanz et al., 2017).

12.3.3 Conception of the empirical survey

In view of the complexity and breadth of the issues and challenges associated with the energy transition in Germany (Czada and Radke, 2019), the particular challenge was to develop a conceptual umbrella and a specific research focus that would allow indicators to be developed across the various sectors of energy consumption (electricity, heat, and mobility), across the most important areas of reform being discussed currently and the three dimensions of social sustainability. The approach was both bottom-up, involving stakeholders and actors from the research contexts, and top-down, involving scientific expertise and concepts. The indicators for the household survey were developed in several steps as outlined as followed:

First, a comprehensive literature review on the topic and a sociologically oriented discourse analysis (Schwab-Trapp, 2011) of the political discourse on the social dimensions of the energy transition were conducted to identify the relevant aspects for data collection. Press releases, position papers, statements, event reports, articles, and reports in print and online media were scanned with regard to which positions, problems, criticisms, demands, conflicts, needs for action or proposals for action were articulated by the "discursive elites" (Schwab-Trapp, 2011: 394) from politics, business, science, civil society, and the media. Both, the literature evaluation and the discourse analysis are continuously followed up as a basis for the further development and adaptation of the survey instrument.

Second, the three dimensions of social sustainability were used to develop overarching sub-areas and corresponding research questions which formed the foundation for developing the indicators (Table 12.2).

TABLE 12.2 The basic research questions of the social sustainability barometer.

Areas	Research questions
General assessment of the Energiewende	How high is the general agreement on the energy system transformation and its goals?
Evaluation of the implementation of the Energiewende	Which concrete measures or developments in central areas of action of the energy system transformation are (very) critically viewed?
Preferences for designing the Energiewende	In the opinion of the population, what should happen in the various fields of action? What do people think is fair?
Evaluation of the federal government and political parties	How strongly do people feel politically represented when questions of the energy transition are concerned? How satisfied are they with the German government's performance in managing the transformation process? Which aspects are viewed particularly critically?
Experiences and expectations with regard to the energy transformation	How is the population affected by the energy system transformation? Which population groups have already had negative experiences?
Participation and willingness to participate	To what extent is the population already actively involved in promoting the energy transition via investments behavior? Can obstacles and potentials be identified?

Third, individual questions or groups of questions were developed in the six sub-areas that were designed to match each indicator for the collection of opinions, attitudes, feelings of justice, experiences, and expectations of the population. The development of the indicators and questions was inspired by the ongoing political debate and the context variables that were typical for the German situation. Since, for example, controversies surrounding cost and distribution justice played an important role in the political discourse of the last years, we adopted several questions to represent this phenomenon in various formats in the questionnaire. The focus of the survey lies on assessing the attitudes toward the goals of the energy system transformation and their implementation in reality. This was accompanied by questions about the perceived performance of the political parties and the federal government as well as the extent and willingness of participation and involvement in energy-related behavior. The price reform and in particular the pricing of CO_2 were included as a current topic in 2018 since this topic has been a hot issue in the debate about reducing the fossil burden from the energy generation. Due to increasing conflicts and protests among the population, onshore wind expansion was also a topic that received wide coverage in the questionnaire. In addition, digitalization of the energy system in households was also included in the surveys. Digital tools may play an important role for the flexibilization of the energy system but it meets many problems such as security of data and protection of privacy. The survey asked for the judgments of the population on the risks and benefits of the upcoming smart meter roll out and the use of flexible electricity tariffs.

On the basis of this approach, the surveys included and integrated the various topical areas of the energy system transformation and provided a database for a systematic and comprehensive interpretation of the present situation with respect to the three dimensions of social sustainability.

To record changes over time, a fixed set of questions is collected either annually or continuously at intervals of 2 or 3 years (monitoring variables). This allows the team to study trends or new developments over time. In addition, there are variables that are collected once or selectively, either because they provide a deeper insight into a specific topic such as the question of how to deal with the challenges of digitalization, promise important additional orientation for the interpretation of the data or contain supplementary questions on specific current topics. This flexibility in the preparation of the questionnaire makes it possible to take adequate account of current developments in society. The standardized questionnaires on which the survey is based can be found at www.iass-potsdam.de/en/barometer.

Some of the main findings of the barometer surveys in the years 2017 and 2018 are presented below as an overview (for more detailed numbers, see Setton, 2019a,b; Setton et al., 2017).

12.4 Key results of the first two barometer surveys

12.4.1 Overwhelming approval for the energy transition

Germans are united in their support for the "Energiewende". The overwhelming majority (90%) of the population in Germany support the energy transition across all educational, income, and age groups, mostly independent of party affiliation, both in rural and urban areas. In addition, the number of people who see the Energiewende as a collective undertaking (80%) to which they themselves and everybody else should contribute is growing perceptibly. Politicians can continue to count on broad public support for a resolute implementation of the energy transition.

12.4.2 Support for all energy transition and renewable objectives

All of the energy policy objectives of the Energiewende enjoy high approval ratings. The frontrunners are increasing energy efficiency (84%), expanding renewable energies (82%), and reducing energy consumption (80%). As the nuclear phase-out, the coal-phase-out is broadly acknowledged as a pillar of Germany's energy policy. The situation is slightly more complicated in the regions and states affected by the phase-out: while the majority of people here are also in favor of the coal exit, the level of skepticism toward the phase-out is far higher than the state and national average. And opposition to the phase-out has grown slightly since 2017. In Germany's second largest lignite region, Lusatia, a relative majority is against the coal-phase-out.

E-mobility is often seen as the key to achieving a climate-friendly mobility transition. But the population has not yet embraced it wholeheartedly. While most people, including car owners, are generally in favor of e-mobility (55%), its approval rating is lower than that of other energy transition goals. People without their own car are more open to e-mobility (64%). A clear majority (53%) is against an exit from the combustion engine by 2030. The younger generation's reluctance to abandon the combustion engine is also astonishing. These results show that far more needs to be done to make e-mobility a more attractive mobility option for people in their everyday lives.

12.4.3 Implementation: criticism of slow progress on climate policies and on disregard of social justice

Although the approval ratings for the energy transition are at a very high level, the population's assessment of its implementation is much more sober. The population's overall assessment of the Energiewende implementation process with regard to costs, political management, citizen-orientation, and fairness is overall mostly negative and significantly more negative than in 2017. Two-third of the population (67%) agree with the statement that the costs of the energy transition are borne by the poor, while the wealthier and the companies profit from it.

Across the political spectrum, more than half of the population is dissatisfied with the Energiewende policies currently being pursued by the German Federal Government. Overall, public confidence in the ability of Germany's political leaders to manage the Energiewende effectively is low and has fallen since the first survey in 2017—the coalition parties have fared particularly badly here. Bündnis 90/the Greens inspire by far the most confidence in this regard.

The reasons cited for this dissatisfaction are revealing: most Germans are concerned that too little progress is being made in the area of climate protection and critical of a social imbalance in the distribution of the costs and benefits of the Energiewende. While a large majority believes that the Energiewende is expensive, most respondents do not cite excessive costs as a major reason for their dissatisfaction. Climate protection and social justice seem to be more important to the population as a whole. What is striking is the different weighting of these two preferences for climate protection and social justice by different population groups (East/West and income group). Both issues are equally important for the further implementation of the Energiewende.

12.4.4 Support for the climate goals—but wish for social balancing

The vast majority of Germans—including the people in lignite-mining states—supports the country's 2020 climate protection goals. But opinions are divided on the question of the importance of climate protection in relation to other sociopolitical objectives (like, for example, safeguarding jobs). Just over one-third of the population believes that rapid climate protection should be prioritized over other issues. At the same time, most Germans feel that it makes sense to slow down the pace of climate protection to accommodate the social needs of the regions. In a nutshell: the majority of Germans wants climate protection and social justice in equal measure. It follows that in the implementation of climate protection measures, the disadvantages for specific population groups and regions should be kept to a minimum and offset wherever possible.

12.4.5 Social balance is very important

The Germans see the state as responsible for a socially just design of the energy system transformation in the state. It should ensure that also low-income earners enjoy sufficient supply of energy. A socially just solution, represented among other things by a progressive tariff, is important to people in Germany. They believe that the onus is on the state to ensure that lower-income households can receive the energy services they need at an affordable price.

12.4.6 CO_2-pricing and fairness in cost distribution

More than two-thirds of the German population is against the exemption of energy-intensive industries from the renewable energy levy. They believe it is wrong that bulk buyers of electricity pay less per kilowatt hour than they do themselves or than other businesses with lower energy demands. The federal government's justification that this is necessary for reasons of competitiveness is equally unpopular among respondents.

Most respondents believe that households and companies with high consumption and high CO_2 emissions should actually make a much higher financial contribution than those who consume less. In addition, the majority believe that a fair distribution of costs means that the more a person consumes, the more he or she should pay per unit consumed. This would ease the burden on poorer households and act as an incentive to higher-income households to consume less energy. This desire for fairness is also reflected in the conviction that those who cause high CO_2 emissions should also bear a large share of the Energiewende costs.

Nowadays, many people in associations, organizations, science, and politics are calling for the introduction of carbon pricing as a central instrument for achieving Germany's climate goals. But for all their good intentions, most people are not prepared to pay more for climate protection in their everyday activities. This attitude is not limited to low-income households and is particularly common among the middle classes. Under these circumstances, it seems likely that the introduction of a carbon pricing system will only be accepted by the broad majority if it is accompanied by a convincing and transparent compensation mechanism. It is, however, just as important to provide people with practicable, affordable, and accessible alternatives to using fossil fuels in their daily lives.

12.4.7 Solar roof panels are by far the most popular renewable technology

The Federal Government's efforts to expand renewable energies in Germany are focused on wind and solar energy. However, public acceptance levels for the different wind and solar technologies vary considerably. Solar energy and solar roof panels in particular continue to enjoy the highest approval rating among Germans, even in densely populated areas. Support is also strong for ground-mounted PV systems, but still significantly lower than that for rooftop solar systems.

In contrast, people are far more skeptical about the expansion of onshore wind energy. Onshore wind energy is the least favored of all the renewable technologies. Although less people are exposed to wind turbines in the vicinity of their homes than to solar roof panels, public acceptance of the former is still lower. So from the point of view of acceptance, it would make sense to concentrate more on expanding rooftop solar systems, which are unlikely to provoke public protests.

12.4.8 Wind expansion: more political participation desired

The majority of people who live in an area with wind turbines are not (particularly) bothered by them, even when they are erected close to their homes. However, a large accumulation of wind turbines on one's doorstep lowers acceptance levels significantly, since the more people adversely affected by the expansion of wind energy, the higher the proportion of the local population that feels inconvenienced. This goes hand in hand with an increasingly negative attitude toward the expansion of onshore wind energy, a growing perception that the Energiewende is injust, and, not surprisingly, a greater willingness to join protests against the erection of new wind turbines. So, the number of wind turbines in one's locality seems to be more decisive for acceptance levels than the distance of the turbines from one's house.

A slim majority of the German population supports the expansion of onshore wind energy but an equally slim majority believes that this expansion should not take place over the heads of the local people who are going to be affected, even if that delays the expansion process. Thus the "wind issue" is at heart a question of democracy; it is about who has a say in the political process. Wind energy imposed from above is likely to meet with growing disapproval. Attempts to increase the financial participation of local communities and citizens in wind energy projects are a step in the right direction.

However, for greater acceptance, there is no way around broad, timely, and well-structured citizen participation, where people have opportunities to influence what happens in their locality.

12.4.9 Digitalization: data protection issues addressed partly—who benefits?

In the important future field of the digitization of the energy system transformation, the population's concerns about data protection and data security are widely spread and have only been partly addressed within legislation. With the roll-out of intelligent metering systems in private households as well, there is a lack of adequate communication with the population and regulatory gaps remain. While the majority rejects to bear the additional costs of the smart-meter roll out, the governments approach place the burden particularly on private households. While most of the people support the smart-meter rollout expecting new opportunities for energy and cost reductions the legislation does by no means guarantee or even make it probable that benefits will be realized.

12.4.10 Little interest in flexible tariffs

Germans admit to being very flexible when it comes to using their household appliances, in particular washing machines and dishwashers. However, large sections of the population do not want to adjust their energy consumption depending on electricity prices or tariff models, or to control it or have it controlled from outside with the help of digital applications. The majority prefers the reliability of fixed tariffs. At the same time, the "traffic-light model," where consumers would be given information on electricity consumption and electricity prices and left to decide whether or not they should adjust their usage, enjoys high approval rates. The low level of interest in flexible tariffs shows that the much-vaunted advantages of a new digital energy world in the home (cost savings, etc.) has only been seized to small degree by private customers. It can therefore be safely assumed that people will not switch to flexible tariffs of their own accord. The diffusion of such tariffs will probably take longer than many scenarios envisage.

12.4.11 Limited enthusiasm for direct participation in new energy systems

People in Germany are aware that the Energiewende also requires their active involvement. There is a willingness to contribute to the energy transition in all sections of the population. The vast majority (86%) of all Germans welcome the fact that citizens can participate in the energy system transformation as energy producers (prosumers). But this involvement still tends to be confined to the purchase of energy-saving household devices and good intentions, for example, to save energy, and very few people are prepared to make substantial financial contributions. So far, only a minority (8%) has invested in intelligent heating control systems and in their own renewable energy systems (10%)—jointly or alone. And the people who have done so are generally homeowners. For most Germans in rented accommodation, this is not an option because, as tenants, they are not in a position to decide on such matters. The willingness to make investments that require additional dedication and knowledge is low, also among many homeowners. Many are reluctant to invest in renewable energy systems or digital heating control. In some cases, this is due to a lack of information, but people are also unsure about what is economical for their own household, which systems are suitable, and how they can protect themselves against data abuse arising from digital technology. There are no signs of a change in investment behavior in the near future. In essence, people are willing to become agents for energy transformations but at this point the majority of the people is reluctant to go beyond low-cost and minor behavioral changes.

12.4.12 The energy transition is a provision for the future: only few expect advantages in the short-to-medium term

Intergenerational justice plays an important role in the population's assessment of the Energiewende, which is seen first and foremost as an investment in a worthwhile future. Only a few people believe that they will feel the benefits of the Energiewende in the short-to-medium term. And many expect it to have a low impact on their personal environment. Indeed, negative expectations prevail when people consider the impact on their financial and economic situation. The perception that the Energiewende is at quite a remove from one's own daily life may explain why large sections of the population do not believe they have many opportunities to contribute and instead rely on politicians to implement this project.

12.5 Conclusion

Comparing the findings presented here as an overwiev with the current policy and the regulatory framework in Germany reveals that the attitudes, experiences, expectations, preferences and concerns of the population are only taken into account to a limited extent in government action. While general approval of the energy system transformation is very high, the implementation process is met with criticism and scepticism by the majority. In addition, trust in political parties and the government is low, and the active involvement of the population needs to be much more supported. In fields of action such as onshore wind expansion there is a mismatch between the regulatory framework and the desire for a stronger political voice of the affected population. In areas such as digitalization of the energy transition and the reform of price components (CO_2 pricing), there is evidence that, despite a general acceptance of the introduction of policies introduced by the government, additional action is needed to ensure maintained acceptance when policies take effect. On the basis of this analysis, more detailed needs for action and options for action can be derived (Setton et al., 2017; Setton, 2019a,b). In sum, there is still a great need for action to make the energy system transformation in Germany socially sustainable. The German Federal Government is well advised to offer more inclusive methods for integrating citizen preferences and values into energy planning and climate policy design. This underlines the main idea of the entire volume that a successful transition requires more inclusive methods and procedures for governance.

The application-oriented concept of social sustainability presented here aims at contributing to the necessary stronger integration of social and ecological aspects with regard to the major social challenges of our time. On the basis of the concept of the social sustainability of the energy transitions, an important knowledge base is provided for political decision-making processes to shape the socio-economic transformation. It indicates where political attention is needed and where perceived deficits need to be addressed. To put the population at the center of social sustainability does not mean to use majority opinions as the sole yardstick for evaluating measures or developments or to rely solely on public preferences. It does, however, mean taking them seriously as an indispensable yardstick and thus taking into account one of the central prerequisites for a successful socio-ecological transformation.

The concept of a socially sustainable energy system transformation has been developed in direct reference to the German Energiewende. In other countries, however, ttransitions to a low-carbon economy are also on the agenda and similar challenges, opportunities, and obstacles have occurred or will occur. It will be important to investigate and integrate social sustainability considerations when designing socio-technical energy systems change. To this extent, the concept can also be applied internationally to other countries and regions. It is important, however, to develop an indicator system specific to the situation of each country.

References

BMUB/UBA - Bundesministerium für Umwelt, Naturschutz, Bau und Reaktorsicherheit/ Umweltbundesamt (Ed.), 2017. Umweltbewusstsein in Deutschland 2016. Ergebnisse einer repräsentativen Bevölkerungsumfrage. Berlin, Dessau.

BMWi – Bundesministerium für Wirtschaft und Energie, Die Energiewende: unsere Erfolgsgeschichte, 2017, Berlin.

BMWi - Federal Ministry of Economics and Energy, 2019. Second progress report "Energy of the Future." June 2019, Berlin.

Brand, K.-W., 2017. Einleitung – Problemstellung und Untersuchungsperspektive. In: Brand, K.-W. (Ed.), Die sozial-ökologische Transformation der Welt, IIn: Campus Verlag, Frankfurt am Main, pp. 13–31.

Czada, R., 2014. Gesellschaft, Staat und Politische Ökonomie im postfossilen Zeitalter. In: Fürst, D., Bache, A. (Eds.), Postfosssile Gesellschaft - Fluchtlinien in die Zukunft (Series: City and Region as Field of Action, Vol. 12. Peter Lang, Frankfurt/M, pp. 13–26.

Czada, R., Radke, J., 2018. Governance langfristiger Transformationsprozesse. Der Sonderfall "Energiewende". In: Radtke, J., Kersting, N. (Eds.), Energiewende: Politikwissenschaftliche Perspektiven. Springer VS, Wiesbaden.

D'Agostino, A.L., Sovacool, B.K., Trott, K., Regalado Ramos, C., Saleem, S., Ong, Y., 2011. What's the state of energy studies research?: A content analysis of three leading journals from 1999 to 2008. Energy 36 (2011), 508–519.

Ekardt, F., 2016. Theorie der Nachhaltigkeit. Ethische, rechtliche, politische und transformative Zugänge - am Beispiel von Klimawandel, Ressourcenknappheit und Welthandel, 2nd, completely revised and updated edition 2016. Baden-Baden: Nomos.

Empacher, C., Wehling, P., 2002. Die sozialen Dimensionen von Nachhaltigkeit. Theoretische Grundlagen und Indikatoren. ISOE study text no. 11. Frankfurt am Main.

Ethics Commission for Safe Energy Supply of the Federal Government, 2011. Deutschlands Energiewende - Ein Gemeinschaftswerk für die Zukunft, Berlin.

Fraune C., Knodt M., Gölz S. and Langer K., (Eds.), 2019. Akzeptanz und politische Partizipation in der Energietransformation, In: Gesellschaftliche Herausforderungen jenseits vonTechnik und Ressourcenausstattung 2019, Springer Fachmedien, Wiesbaden.

Gabriel, O., Völkl, K., 2008. Politische und soziale Partizipation. In: Gabriel, Oscar W. and Kropp, S. (Eds.), Die EU-Staaten im Vergleich. Strukturen, Prozesse, Politikinhalte, 2008, Wiesbaden: VS Verlag für Sozialwissenschaften, pp 269–298.

Gaede, J., Rowlands, I.H., 2018. Visualizing social acceptance research. A bibliometric review of the social acceptance literature for energy technology and fuels. Energy Res. & Soc. Sci. 49 (2018), 142–158.

Geels, F.W., Sovacool, B.K., Schwanen, T., Sorrell, S., 2017a. Sociotechnical transitions for deep decarbonization. Science 357 (6357), 1242–1244.

Geels, F.W., Sovacool, B.K., Swanen, T., Sorrell, S., 2017b. The socio-technical dynamics of low-carbon transitions. Joule 1 (3), 463–479.

Grasselt, Nico, 2016. Die Entzauberung der Energiewende. Politik- und Diskurswandel unter schwarz-gelben Argumentationsmustern. Springer VS, Wiesbaden.

Grießhammer, R., Brohmann, B., 2015. How Transformations and Social Innovations Can Succeed. Transformation Strategies and Models of Change for Transition to a Sustainable Society. Nomos, Baden-Baden.

Groh, E.D., Ziegler, A., 2017. On self-interested preferences for burden sharing rules: an econometric analysis for the costs of energy policy measures. Joint Discussion Paper Series in Economics by the Universities of Aachen, Gießen, Göttingen, Kassel, Marburg, Siegen. No. 54-2017. Marburg.

Grunwald, A., Renn, O., Schippl, J., 2018. Die Energiewende verstehen – orientieren – gestalten: der Ansatz der Helmholtz-Allianz ENERGY-TRANS. In: Holstenkamp, L., Radtke, J. (Eds.), Handbuch Energiewende und Partizipation. Springer VS, Wiesbaden, pp. 829–846.

Heyen, D.A. and Brohmann, B., 2017. Konzepte grundlegenden gesellschaftlichen Wandels und seiner Gestaltung Richtung Nachhaltigkeit – ein Überblick über die aktuelle Transformationsliteratur, In: Rückert-John, Jana (Ed.), Governance für eine Gesellschaftstransformation, Innovation und Gesellschaft, Springer VS, Wiesbaden, 69–86.

Holstenkamp, L., Radtke, J. (Eds.), 2018. Handbuch Energiewende und Partizipation. Springer VS, Wiesbaden.

Holtmann, E., 2015. Die Entwicklung der Demokratie. Legitimationsverlust und Reformbedarf? In: Harles, L. and Lange, D. (Eds.). Zeitalter der Partizipation. Paradigmenwechsel in Politik und politischer Bildung? Schriftenreihe der DVPB. 2015, Frankfurt/M, Wochenschau Verlag, 63–73.

IPCC, 2018. Special Report: Global Warming of 1.5 ọ C. October 2018.

Jahrbuch Ökologische Ökonomik, 2007. Soziale Nachhaltigkeit. Vol. 5., 2007, Metropolis, Marburg.

Kopfmüller, J., Brandl, V., Jörissen, J., Paetau, M., Banse, G., Coenen, R., 2001. Nachhaltige Entwicklung integrativ betrachtet. Konstitutive Elemente, Regeln, Indikatoren. edition sigma, Berlin.

Littig, B., Grießler, E., 2005. Social sustainability: a catchword between political pragmatism and social theory. Int. J. Sustain. Dev. 8 (Nos. 1/2), 65–79. 2005.

Löschel, A., Erdmann, G., Staiß, F., Ziesing, H.-J., 2019. Expertenkommission zum Monitoringprozess "Energie der Zukunft," Stellungnahme zum zweiten Fortschrittsbericht der Bundesregierung für das Berichtsjahr 2017. Berlin, Münster, Stuttgart. May 2019.

Matuschke, I., Renn, O., 2019. Bürgergutachten zur gerechten Verteilung von Stromkosten. In: Setton, D., Soziale Nachhaltigkeit wagen. Umfassende Auswertung der Daten des Sozialen Nachhaltigkeitsbarometers der Energiewende 2017/2018. IASS-Study, December 2019.

Michelsen, G. and Adomßent, M., 2014. Nachhaltige Entwicklung: Hintergründe und Zusammenhänge. In: Heinrichs, H. and Michelsen G. (Eds.): Nachhaltigkeitswissenschaften, Springer-Verlag, Berlin/Heidelberg. 3–59.

Miller, C.A., et al., 2015. Socio-energy systems design: a policy framework for energy transitions. Energy Res. Soc. Sci. 6, 29–40.

Miller, C.A., Iles, A., Jones, C.F., 2013. The social dimensions of energy transitions. Sci. Cult. 22/2, 135–148.

Miller, C.A., Richter, J., 2014. Social planning for energy transitions. Curr. Sustain. Renew. Energy Rep 2014 (1), 77–84.

Nanz P., Renn O. and Laurence M., 2017. The transdisciplinary approach of the Institute for Adcanced Sustainability Studies (IASS): concept and implementation, Gaia Ec.o. Perspect. for Sci. and Soc. 23/6, 293–296.

Opielka, M., 2017. Soziale Nachhaltigkeit. Auf dem Weg zur Internationalisierungsgesellschaft, Oekom, Munich.

Renn, O. (Ed.), 2015a. Aspekte der Energiewende aus sozialwissenschaftlicher Perspektive (Analysis from the series Energiesysteme der Zukunft), Munich.

Renn, O., 2015b. Akzeptanz und Energiewende. Bürgerbeteiligung als Voraussetzung für gelingende Transformationsprozesse. In: Jahrbuch für Christliche Sozialwissenschaften JCSW 56, 133–154

Renn, O., Dreyer, M., 2013. Risiken der Energiewende: Möglichkeiten der Risikosteuerung mithilfe eines Risk-Governance-Ansatzes. In: Vierteljahreshefte zur Wirtschaftsforschung, 82, DIW Berlin.

Renn, O., Deutschle, J., Jäger, A., Weimar-Jehle, W., Leitbild Nachhaltigkeit. 2007. Eine normativ-funktionale Konzeption und ihre Umsetzung, Springer VS, Wiesbaden.

Ritt, T. (Ed.), 2002. Social Sustainability: From Environmental Policy to Sustainability? Federal Chamber of Labour, Vienna.

Rösch, C., Bräutigam, K.-R., Kopfmüller, J., Stelzer, V., Lichtner, P., 2017. Indicator system for the sustainability assessment of the German energy system and its transition. Sustainability Soc. 2017 (7), 1.

Saidi, A., 2018. Einstellungen zur Energiewende in Norddeutschland. Erste Befragung im Rahmen der Akzeptanzforschung für das Projekt NEW 4.0; Universität Hamburg.

Schippl, J., Grunwald, A., Renn, O., 2017. Die Energiewende verstehen – orientieren – gestalten. Einsichten aus fünf Jahren integrativer Forschung. In: Schippl, J., Grunwald, A., Renn, O. (Eds.), Die Energiewende verstehen – orientieren – gestalten: Erkenntnisse aus der Helmholtz-Allianz ENERGY-TRANS. Nomos, Baden-Baden, pp. 9–34.

Schumann, D., 2017. Public perception of energy systems transformation in Germany. Euro-Asian J. Sustain. Energy Develop. Policy 5/2, 33–56.

Schwab-Trapp, M., 2011. Diskurse als sozialogisches. In: Konzept, Keller, R., Hirseland, A., Schneider, W., Viehöver, W. (Eds.), Handbuch Sozialwissenschaftliche Diskursanalyse: Volume 1: Theories and Methods. Springer VS, Wiesbaden, pp. 283–307.

Setton, D., 2019a. Soziale Nachhaltigkeit wagen. Die Energiewende aus Sicht der Bevölkerung. Umfassende Auswertung der Daten des Sozialen Nachhaltigkeitsbarometers der Energiewende 2017/2018. 2019, IASS-Study, December 2019.

Setton, D., 2019b. Social Sustainability Barometer for the German Energiewende: 2018 Edition. Core statements and summary of the key findings, IASS Study, July 2019.

Setton, D., Matuschke, I., Renn, O., 2017. Social Sustainability Barometer for the German Energiewende 2017: Core statements and summary of the key findings. IASS Study, November 2017.

Sonnberger, M., Ruddat, M., 2016. Die gesellschaftliche Wahrnehmung der Energiewende – Ergebnisse einer deutschlandweiten Repräsentativbefragung. In: Stuttgarter Beiträge zur Risiko- und Nachhaltigkeitsforschung, (Stuttgart Contributions to Risk and SustainabilityResearch), No. 34.

Sovacool, B.K., 2014a. What are we doing here? Analyzing fifteen years of energy scholarship and proposing a social science research agenda. Energy Res.Soc. Sci. 1, 1–29.

Sovacool, B.K., 2014b. Review. Exposing the paradoxes of climate and energy governance. Int. Stud. Rev. 2014 (16), 294–297.

Upham, P., Oltra, C., Boso, À., 2015. Social acceptance of energy technologies, infrastructures and applications: towards a general cross-paradigmatic analytical framework. Energy Res. Soc. Sci. 8, 100–112.

WBGU - German Advisory Council on Global Change, 2011. World in Transition: A social Contract for sustainability. WBGU, Berlin.

WCED - World Commission on Environment and Development, Brundtland Report, 1987. Our common future, United Nations, Oxford University Press, Oxford, New York.

Wuppertal Institute, 2016. Knowledge as Transformative Energy. For linking models and experiments in the building energy system transformation. Oekom, Munich.

Wüstenhagen, R., Wolinsk, M., Bürer, M.J., 2007. Social acceptance of renewable energy innovation: An introduction to the concept. Energy Policy 35, 2683–2691.

Ziegler, A., 2017. Economic calculus or personal and social values? A micro-econometric analysis of the acceptance of climate and energy policy measures. Joint Discussion Paper Series in Economics by the Universities of Aachen, Gießen, Göttingen, Kassel, Marburg, Siegen. No. 16-2017. Marburg.

CHAPTER

13

Conclusion

Ortwin Renn[1], Anna Deckert[2] and Frank Ulmer[2]
[1]Institute for Advanced Sustainability Studies (IASS), Berliner Strasse, Potsdam, Germany
[2]Dialogik Non-profit Institute for Communication and cooperation research, Stuttgart, Germany

The objective of this volume was to address and discuss the need and potential for stakeholder involvement and public participation in designing sustainable energy policies. In the first part of the book, Renn and Schweizer introduced different conceptual models of how to respond to the two crucial questions of any involvement process: Who and what should be included? What kind of output should the involvement process produce in order to facilitate better decision-making? Furthermore, this chapter explained the combination of analytic and deliberative components when approaching complex, uncertain, and ambiguous energy challenges. Such wicked energy problems require epistemic, reflective, and participatory discourse structures. For each of these discourse types, appropriate formats were suggested and explained.

For making decisions in a complex energy context, a combination of functionalist (analytic) and deliberative perspective has been proven most effective even in complicated risk debates. The US-National Academy of Sciences has recommended the analytic–deliberative discourse as the most appropriate form of risk decision-making (Stern and Fineberg, 1996). If the issue at hand is about how to resolve complexity an epistemic, knowledge-based discourse is recommended with the respective formats of expert workshops, Delphi procedure, scientific consensus conferences, and other knowledge-driven methods. In situations of high uncertainty, questions of equity and fairness become paramount. Policies need to address fair benefit and risk sharing. The objective is to find a fair compromise between those who will benefit from the policy (such as increasing the price for fossil fuel through higher taxation) and those who may suffer from the risks. This can be done in a reflective discourse including formats such as Roundtables, mediation, and value tree exercises. Finally, if the debate centers arround values that are affected by energy policies, it is essential to include public preferences and concerns. Neither agency staff nor scientific advisory groups are able or legitimized to represent the full scope of public preferences and values. This is a compelling reason for broadening the basis of decision-making and

The Role of Public Participation in Energy Transitions
DOI: https://doi.org/10.1016/B978-0-12-819515-4.00013-1

including those who have to "pay" in terms of bearing the consequences of far reaching energy decisions. Formats that fit into the participatory discourse type are focus groups, citizen advisory panels, consensus conferences, and citizen panels (juries).

Organizing and structuring stakeholder involvement and participation projects require excellent communication and organizational skills but also sufficient experience in handling conflicts and facilitating discourses. Public participation in this contested field goes beyond the well-meant intention of having stakeholders involved in expressing their preferences. The mere desire to initiate a two-way communication process and the willingness to listen to public concerns are insufficient. Decisions for more sustainable energy systems must reflect effective regulation, efficient use of resources, legitimate means of action, and social acceptability. The combination of analytic rigor and deliberative democracy is the best medicine for dealing with these complex issues.

The objective is to find an organizational structure and procedure so that all elements of the selected model contribute to the deliberation process the type of expertise and knowledge that claim legitimacy within a rational and fair decision-making procedure. The book includes many examples of how to design such processes, for example, for siting of wind farms, for getting residents involved in new business arrangements for solar energy (prosumer), or in assisting local government in planning renewable energy generation. Each case demonstrated that there is no recipe for what works, but each approach to include stakeholders and citizens demands a thorough analysis of the conditions and history that characterize the location and to design the right format in line with the six concepts developed in the theoretical part. The participation process needs to be carefully designed and evaluated. It does not make sense to replace technical expertise with vague public perceptions; nor is it justified to have the experts insert their own value judgments into what ought to be a democratic process.

Recently, there has been much concern among energy policy-makers that opening the risk management arena to stakeholder input would lead to a dismissal of actual knowledge and to inefficient spending of public money (Sunstein, 2002). Given the experience with stakeholder involvement that have been documented in this book and elsewhere, these concerns are not warranted. There are only a few voices that wish to restrict scientific input to improve policy-making for designing sustainable energy practices. Scientific expertise is an essential element of stakeholder involvement and a crucial pillar of the analytic–deliberative concept. The role of scientific analysis in energy policy-making should not be weakened but, rather, strengthened when opening the arena for stakeholder input. Profound scientific knowledge is required in any type of governance, especially with regard to dealing with complexity. This knowledge has to be assessed and collected by scientists and experts, who are recognized as competent authorities in the respective risk field. The systematic search for the "state of the art" in energy supply and demand leads to a knowledge base that provides the data for deliberation. At the same time, however, the style of deliberation also should transform the scientific discourse and lead the discussion toward classifying knowledge claims, characterizing uncertainties, exploring the range of alternative options, and acknowledging the limits of systematic knowledge in many energy applications.

Many arguments in favor of analytic–deliberative processes and their theoretical foundations provide ample evidence for their potential contribution to improving risk evaluation and management (Rauschmayer and Wittmer, 2006). It is still an open question whether stakeholder involvement processes can deliver what they promise in theory. The empirical account is still open and incomplete. Being active in developing and testing analytic–deliberative processes, one can be confident, however, that over time we will not just prove, theoretically the merits and potentials but also confirm the practical feasibility and superiority of analytic–deliberative processes in different political cultures and among a variety of regulatory styles.

But even successful stakeholder dialogs do not automatically lead to greater acceptance, more infrastructure and thus to a faster energy system transformation. The key issue for a successful energy system transformation will be how to strengthen connectivity to political processes. Without the systematic examination of results from stakeholder dialogs and participation processes and a subsequent courageous examination and action by politicians, the elaboration of high-quality results as part of the participation processes seems pointless. This connectivity to policy-making can be achieved through various measures, which go beyond the topic of this book, like strengthening of administration in being able to react agilely to change requests from politics. Further research should be conducted into the conditions under which participation and subsequent policy advice lead to political and social action.

This book, nevertheless, demonstrates that at least for the German energy transition, citizen participation and stakeholder involvement have been crucial for progressing on the route to sustainable energy systems and will be instrumental for reaching the next steps.

References

Rauschmayer, F., Wittmer, H., 2006. Evaluating deliberative and analytical methods for the resolution of environmental conflicts. Land Use Pol. 23 (2006), 108–122.

Stern, P.C., Fineberg, H.V., 1996. Understanding Risk: Informing Decisions in a Democratic Society. National Academy Press, Washington, DC.

Sunstein, C., 2002. Risk and Reason. Cambridge University Press, Cambridge.

Index

Note: Page numbers followed by "*f*" and "*t*" refer to figures and tables, respectively.

F

R

S